William Lewis Baker

The Beam

Or technical elements of girder construction

William Lewis Baker

The Beam
Or technical elements of girder construction

ISBN/EAN: 9783743687165

Printed in Europe, USA, Canada, Australia, Japan

Cover: Foto ©berggeist007 / pixelio.de

More available books at **www.hansebooks.com**

THE BEAM

OR

TECHNICAL ELEMENTS OF
GIRDER CONSTRUCTION

BY

WILLIAM LEWIS BAKER, A.M.I.C.E.

LONDON: CHAPMAN & HALL, L^{D.}

———

1892

PREFACE.

THE subject of the following pages is treated from the foundation, with the object of making it useful to the student who may possess but little previous acquaintance with the subject itself, or with mathematics, or the theory of mechanics. No complicated algebraic formulæ have been found necessary for the determination of any of the questions which arise, either in respect to the amount of stresses or strains, or to the amounts necessary to provide for them. On the other hand such elementary formulæ as are required for the ordinary routine of constructive work are fully explained.

It has further been attempted, by a series of extremely simple examples, to illustrate the methods which may be followed in the solution of all such cases as will be found to occur in practice, and at the same time to provide easily accessible and reliable data for reference, and for the basis of calculations.

CONTENTS.

CONTENTS.

THE BEAM.

CHAPTER I.

1. A Tie is a bar or piece of any suitable material attached at its extremities to opposite parts of a structure, in order by its tensile strength to resist any tendency such parts may have to separate.

2. A Strut is a piece, or combination of pieces, so placed that by its compressive strength it may transfer a thrust from one part of a structure to some other part, or beyond it.

3. A Beam is a single piece or structure of any material or combination of materials extending longitudinally over space to one or more points of support or resistance, for the purpose of carrying a load or pressure distributed over the whole or over a part of its length, or applied at any point or points in its length. The Deck Beams of a Vessel, the Beam of a Steam Engine, a Scale Beam, or a Weaver's Beam are examples.*

---

* The well-known terms *Tie-beam* and *Collar-beam* are not consistent with the above definition, and, as the office of the former is to resist tension, and the latter to resist compression, each in the direction of its length, it is submitted that *Tie-bar* and *Collar-bar* would be preferable expressions.

B

4.   A Girder is the name originally and still given to the main wooden beams upon which flooring joists are laid, but the term is more especially applied to any beam employed in construction for the purpose of transferring horizontally the pressure or weight of a load from one or more points or surfaces above the space which it crosses to other points upon which it is supported.

The principal varieties of girders may be technically divided into the eight following classes :—

(1.) *The Girder* of three members, consisting of a top and a bottom member or *Table* united by an intermediate vertical member called the *Web*. When of cast iron the upper table is usually small, and the transverse section of the girder resembles an inverted T (Fig. 1), but when the material used is either rolled steel or iron, then the transverse section resembles the letter H, placed sideways (Fig. 2).

Top Table

Web.

Bottom Table

Fig. 1.

Fig. 2.

(2.) *The Box Girder*, usually constructed of rolled steel or iron, is so named from having two webs placed apart from each other, which, together with the tables, enclose a space.

(3.) *The Tubular Girder* is a large structure of the Box Girder type, for railway or other bridges, and so constructed that a rolling or other load may pass through the enclosed space or rectangular tube formed by the tables and webs.

(4.) *The Framed Girder*, in which the web is formed of a simple framework of struts and ties.

(5.) *The Lattice Girder*, in which the web is formed of

four or more series of lattice bars crossing each other diagonally, and secured by rivets or bolts at their extremities to the tables, and also laterally to each other at their crossings.

(6.) *The Bowstring Girder*, in which the upper table is arched upwards like a bow, and the lower resembles the string in tension.

(7.) *The Flitch Girder*, usually formed of two planks of timber with an intermediate steel or iron plate, all of which are placed laterally against each other and bolted together.

(8.) *The Continuous Girder*, extending in one piece or continuous length over two or more spans. An ordinary length of railway rail extending over several chairs is essentially a continuous girder.

5. A Cantilever is a beam or bracket projecting from a wall or other support, and designed to carry a load from or by one support only, and to transfer the weight of the load to such support. Any projecting portion of constructive work does duty as a cantilever in supporting its own weight, and, as the case may be, frequently a load in addition—as for instance a projecting cornice, a balcony, a landing, or stone steps projecting from a wall and otherwise unsupported.

The " hammer beams " of mediæval roofs are essentially cantilevers.

6. A Truss is an open triangulated framework of struts and ties, and a top and bottom member, riveted, bolted, or tenoned and morticed together, as the material used or the nature of the work may require, with a bearing at each end, for the purpose of carrying a load or resisting external weight or pressure.

Trusses are adopted when considerable depth is necessary, and the form of an ordinary girder would consequently be unsuitable, as in the case of roof principals; or when lateral rigidity, of which they have little, is either of minor importance or can be imparted from without—as for instance by the *purlins** of a roof, or by the boarding of a roof or floor.

To Truss is to impart additional strength or rigidity to a beam or structure, or to any part of it, by means of one or more struts and ties so united as to constitute a truss.

ELEMENTARY PRINCIPLES.

7. The Office of the Web of a girder is to transmit the vertical pressure or weight of the load to the supports, or from one of the tables to the other; in so doing the web will resist all *vertical* and *diagonal* stresses induced by the load.

8. The Office of the Tables is to resist all *horizontal* stresses occasioned by the load.

9. A Stress is a force set up within the substance of a structure by the action of its own weight or of an external load or pressure.

All stresses are accompanied by corresponding disturbances of material or strains.

A Strain is a fractional elongation or contraction or other displacement of the particles of a body, which may either

* *Purlins* are the longitudinal members of a roof which support the rafters at intermediate points between their feet and the ridge. Purlins are themselves frequently constructed as trusses. Framed and lattice girders are in fact, trussed beams.

cease with the cause, or may become permanent, wholly or in part, in which case the material is said to have taken a *set*.   Or strain may end in rupture or collapse.

Thus it should not be forgotten in constructive work that the particles of all materials exist in a chronic state of more or less continual strain and consequent relative dis-integration.

There are five conditions of strain in which a solid may be placed: 1 *tensile*, 2 *compressive*, 3 *shearing*, 4 *transverse* or bending, and 5 *torsional* or twisting.

10. Transverse Strength in a beam or girder is the resistance which it is capable of opposing to any force or forces acting upon it in any direction not parallel with its length.

A beam may have to resist stresses caused by either upward or downward transverse pressure, or by both, as in the case of engine beams and crossheads.   A beam may also be placed vertically or in any other position for the purpose of resisting transverse pressure.

11. Relative Position of Beam and Load.   In the following resolutions of the elementary laws which govern strains in beams or girders, the direction of their length is assumed to be horizontal with loads causing downward vertical pressure, these being their usual conditions.

12. The Weight of a Beam may be separately treated as a distributed load, and the stresses caused thereby may then be respectively added to those caused by whatever superimposed load the beam may have to carry.

13. Unit of Length, Area, Weight, Stress, or Strain.   It is usual in calculating, or in what is technically called "taking out," the stresses in, or the strength of beams or girders, to adopt, as may be convenient, the foot or the inch as the unit of lineal, superficial, or cubic measurement, and the ton, cwt., or lb. as the unit of weight or pressure.

The unit of stress most commonly adopted is the pound or ton per square inch. The proper measure of strain is the amount of elongation, or contraction, or other displacement; but, so long as the strain varies with the stress, which is the case within the limits of perfect elasticity, whatever measures the one measures also the other.

In an examination or solution of elementary principles, a quantity may be simply expressed as so many *units* of length, area, weight or pressure, stress or strain.

14. A Moment is, in statics, a technical expression, denoting the measure of the tendency of any force to *turn* a body or system in any direction round a given point, and this measure is the multiple of the force or pressure into the length of the perpendicular drawn from the given point to the line of the direction of the force.

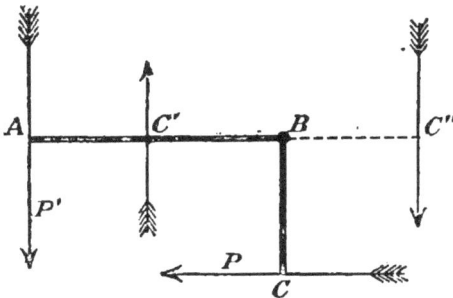

As an example, let B, Fig. 3, be a given point from which draw a perpendicular B C=6 length units to the line of the direction of a pressure P=11 pressure units acting in the direction of the arrow.

Fig 3.

Again, from the point B draw a perpendicular B A =

11 length units to the line of the direction of another pressure $P' = 6$ pressure units also acting in the direction of an arrow.

Then the moment of the pressure P about the point $B = 11 \times 6 = 66$, and the moment of the pressure P' about the same point $= 6 \times 11 = 66$. Therefore the moments of the two pressures P and P' tending to turn the bent lever A B C about the point B are equal, and as the one pressure tends to turn it in an opposite direction to that of the other pressure, and the moment of the one has been found to be equal to that of the other, the system is consequently in a state of equilibrium.

It is further evident that the result would be the same whatever may be the angle A B C contained by the perpendiculars drawn to the lines of the directions of the pressures; or again, if a point $C' = 6$ units from B in the arm A B of the lever were substituted for the arm B C; or further, if the arm B C were in a straight line with A B and the system became the straight lever A B C'', the perpendicular direction of the pressures relatively to the arms being maintained.*

Thus the *principle of the lever* is based on the equality of moments measured from the fulcrum.

* "If any number of pressures act in the same plane and any point be taken in that plane, and perpendiculars be drawn from it upon the direction of all these pressures, produced if necessary, and if the number of units in each pressure be then multiplied by the number of units in the corresponding perpendicular, then this product is called the *moment* of that pressure *about* the point from which the perpendiculars are drawn, and these moments are said to be measured from that point."—*Moseley.*

Thus the sum of the moments of any number of pressures acting in the same direction in the same plane round a given point is equal to the moment of their resultant.

It will have been seen that the forces P and P' are inversely as the lengths of the arms of the lever upon which they act. Thus if the length A B + B C" be divided into as many units as there are units of pressure in P + P', then if equilibrium is to be maintained the position of the fulcrum will be inversely as those pressures measured from each.

A lever may be employed under the three following conditions :

With the fulcrum between the power and the load ;

With the load between the power and the fulcrum ; or,

With the power between the load and the fulcrum.

But in each case the moments of P and P' are measured from the fulcrum.

15. Parallelogram of Pressures. If three pressures acting upon a point are in equilibrium, and a line be drawn from this point in the direction of each pressure, each line being as many units in length as there are units in the relative pressure, then these lines will form the adjacent sides and diagonal of a parallelogram.

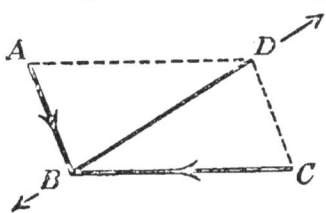

Fig. 4.

Let B, Fig. 4, be a point, and the lines A B and C B represent the directions and amounts of two thrust pressures acting upon the point B, as in‧dicated by the arrow-heads. Draw A D equal to and parallel with B C, and C D equal to and parallel with A B. Then the diagonal B D of the parallelogram, when read in connection with the arrow at B, is the resultant of the two thrust pressures, in respect of both measure and direction. The same line B D, when read in connection with the arrow at D, represents the measure and direction of a reverse pressure or resistance, which would exactly balance the two pressures, or place them in a state of equilibrium.

If A B = C B, and A B C be a right angle, then the resultant B D will be the diagonal of a square a side of which being = 1, B D = $\sqrt{2} = 1.4142.$

If A B and B C were tensile forces, acting upon the point B, then B D would be a thrust resistance and the arrow-heads should be reversed.

16. First Principles of Horizontal Strain in Beams. Let A B, Fig. 5, represent a beam equally loaded on each side of the central point of support c by the weights W W‘. Now if the upper part of this beam were sawn through in any part of its length as at d, and a fulcrum assumed at f, then the vertical pressure caused by the weight W acting through the portion A fd of the beam, as through a bent lever, will tend

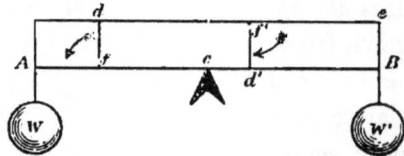

Fig. 5.

to widen horizontally the space at d left there by the saw cut. If, therefore, the saw cut had not been made, the upper part of the beam at d would obviously have been in a state of horizontal tensile strain.

Again, let the lower part of the beam be sufficiently divided or sawn through transversely at any part d', and a fulcrum assumed at f', then the vertical pressure caused by the gravitation of the weight W' acting through the portion B $d' f'$ of the beam will, through the bent lever $e f' d'$, close up horizontally the space at d' left by the saw cut made there. Therefore, with or without the saw cut, the lower part of the beam at d' would evidently be in a state of horizontal compression.

Now, by the law that action and reaction are equal and opposite when equilibrium is maintained, because any part d of the beam is in tension, the opposite part f will be in compression ; and, because a part d' is in compression, then the opposite part f' will be in a state of tension.

Hence the primary vertical pressure of the load subjects the whole of the upper part of the beam to a state of horizontal tension, and at the same time places the whole of the lower part in a state of horizontal compression, and, equilibrium having been assumed, the moments of stress in the two parts must be equal because they are opposite.

17. The Neutral Axis and the Neutral Plane. It follows, then, that in any vertical line drawn in any vertical section of a beam, there must be a point in which horizontal compression and tension vanish and are reversed. Let such a point be assumed in the same plane in any number of vertical sections, then a line drawn through all these points will represent the *neutral axis* of the beam; and further, if we suppose the neutral axis to be extended laterally throughout the breadth of the beam, we shall then obtain what is termed the *neutral plane*, in which both compression and extension cease, and consequently there is no horizontal strain whatever.

18.   The Position of the Neutral Axis. The condition of the beam A B, Fig. 6, supported at each end and subjected to an intermediate pressure or load P, is the reverse of that shown in Fig. 5 ; the strains are therefore reversed, the part below the neutral plane $n'\,n\,n''$ being in tension and the part above it in compression. Now it has been mathematically proved that within the limits of perfect elasticity whatever the section of a beam may be, and whatever difference there may exist in the ultimate amount of resistance its particles or fibres are capable of opposing to tensile as compared with compressive stress, the *neutral axis* ordinarily passes through the *centre of gravity* of any

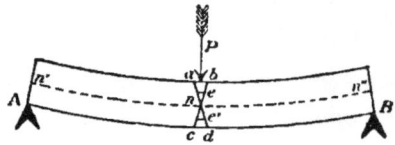

Fig. 6.

given transverse section. Suppose A B, Fig. 6, to be a solid rectangular beam, the neutral axis n' n n'' will thus pass longitudinally along its centre.

From the central point n draw two equal isosceles triangles a b n and n c d with their bases coinciding with the top and bottom of the beam. Then the base a b will represent the maximum horizontal compressive stress, and the base c d the maximum horizontal tensile stress that a transverse section has to resist, while the length of any intermediate lines as e, e', drawn parallel to a b and c d, will represent the proportional amounts of stress at those lines. Now if the two triangles were filled in with a regular series of such lines, the total number of their units of length in the one triangle would equal the total number of their units of length in the other, or the areas of the triangles being equal, the area of the triangle a b n would represent the total amount of compressive, and the area of the triangle n c d the total amount of tensile stress which the section undergoes. The beam would thus be deflected downwards as in the diagram, its fibres or particles above the neutral plane would be longitudinally compressed, while those below it would be longitudinally elongated.

The moments of the resistance of a section of a beam being measured from its neutral axis, which in the case of a solid rectangular beam divides the section into two equal parts, the one part being in compression and the other in tension, the ultimate strength of the section will depend upon the half which from the nature of the material is capable of offering the *least* amount of resistance to either of those stresses.

Thus taking cast iron, the tensile strength of which is less than the resistance it is capable of offering to compression, rupture would first commence by rending at the bottom of the solid rectangular beam (Fig. 6), whereas the reverse being the case with wrought iron, failure would first commence by crushing at the top of the beam. But whether the stress were sufficient to produce rupture or not, there would be an over-plus of strength above the neutral axis in the cast-iron beam, and an overplus of strength below the neutral axis in the wrought-iron beam.

These considerations, therefore, naturally led the way to the more advanced forms of section as shown in Figs 1 and 2, p. 2, in which the ultimate resistances of areas of section to compression and extension may be so adjusted as to be of equal efficiency, at the same time practically bringing within reach the most economical forms in the adaptation of materials to the construction of beams.

19. The Efficiency of the Depth of Beams.
Until fracture takes place or is approximately near, the amount of strain and of resistance to strain, that is stress, per unit of area, in any portion of a section of a beam, whether in exten-sion or compression, will vary as its distance from the neutral axis.

If we suppose the section of any two solid rectangular beams of the same width, but of unequal depths, to be divided into an equal number of small corresponding elements, the *area* of each of these elements will vary as the depth of the beam, and the *distance* of each such element from the neutral axis will also vary in the same proportion. The strength, therefore, of the beam will vary as the square of the depth.

20. The Efficiency of Breadth in Beams. It will have been apparent from what has been already advanced that strain in a beam, and therefore its resistance to strain, takes effect in vertical planes, which may be represented by any lines A B, C D, E F in the cross section Fig. 7, and these planes extend throughout the length of the beam. So that if a beam of uniform breadth were supposed to be formed of a series of plates of a given width placed vertically side by side, then, by increasing or diminishing the number of such plates, we should increase or diminish in the same ratio the strength of the section,

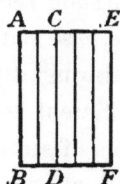

Fig. 7.

and so the strength of a rectangular beam one inch wide would be increased five times by making it five inches wide. The strength of any transverse section of a rectangular beam varies therefore *directly as its breadth.*

21. Vertical or Shearing Strain in Beams. The principles which regulate the effects of this strain are the same in any form of beam. Their consideration, however, with reference to solid beams of uniform section would be quite unnecessary, for when the sectional area of such beams is sufficient to resist the horizontal stress set up by a load, it would be found to be more than sufficient to resist the direct vertical stress of the load; and moreover, the direction of the one being at right angles to that of the other, they in no way act in conjunction or augment each other.

The case of webbed girders is, however, different; for their tables, being simply horizontal plates, are practically unfitted by their form and position to resist vertical stress. Thus the tables having no appreciable vertical rigidity, this stress must devolve upon and be taken by the web But the tables nevertheless impart efficient lateral rigidity to the web.

Referring back to Fig. 5 (**16**), the weight of each load W and W′ must travel through each respective half of the beam to the central support *c*. There is therefore a vertical stress = W at *any* vertical section of the beam between A and *c*, and as W′ = W there is also the same stress between B and *c*.

In a like manner a beam, whether supported at its centre, Fig. 5 (**16**), or at each end, Fig. 6 (**18**), transmits intact the total amount of the vertical pressure of the load direct to the support or supports. This pressure, therefore, or in the case of a distributed load a component fraction of it, will cause an equal stress at any vertical section of a beam, setting up therein what is termed *shearing strain*.

Fig. 8 is given* in illustration of a model provided at the City and Guilds of London Technical College,

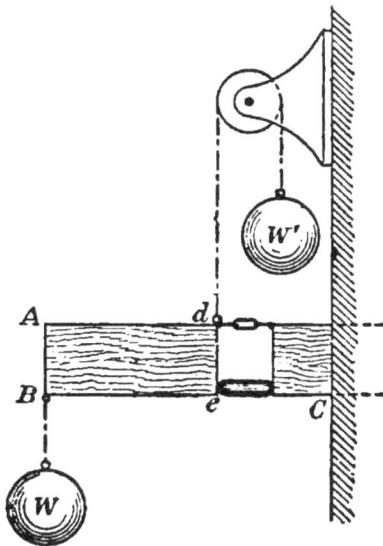

Fig. 8.

* By the kind permission of Professor John Perry, the author of "Practical Mechanics." Cassell and Co.

for the purpose of elucidating the action of vertical or *shearing strain* in beams.

Let A B C represent one-half of such a beam as that shown Fig. 5 (**16**), but with the other half built into a wall, the external half thus forming a cantilever with a weight W suspended at its outer end A B. Let any part of this beam be removed, and in its place let a cord or link which is only capable of resisting tension be inserted at *d*, and a rod or strut which is only capable of transmitting thrust be inserted at *e*, these being the two horizontal stresses which would be set up by the weight W in those parts were the beam intact. It is evident that such an arrangement will not in itself keep the portion A B *d e* in position without an upward force at *d e* equal to the weight W, plus the weight of the severed part A B *d e* of the beam. Thus at any vertical section—as *d e* —of a beam, besides horizontal resistance to tension and compression, resistance to shearing stress is an essential element of strength, and this is supplied in the model by a weight W′ connected with and acting in the direction of the section *d e* by means of a cord passing over a pulley.

Until discharged upon the supports, shearing stress is in practice always assumed to be resisted entirely by the web of a plate girder.

22. Diagonal Strain in Beams. Another important office of the web in addition to transferring the vertical pressure of the load to the points of support, is to transfer to the tables the horizontal stresses due to leverage. Now as it is impossible that vertical stress in a structure can either travel in a lateral direction, or can be resolved directly into horizontal stress, the latter being at right angles with the former, it necessarily follows that both these

operations are effected by diagonal stresses set up in the
web.   That this is so was clearly demonstrated by the experiments made by Mr. Robert Stephenson upon model girders
with reference to his design for the Menai Bridge.   Besides, the buckling of the web plates of these models was
found to demonstrate unequivocally that the diagonal stresses
had a tendency to take the direction of an angle of 45°.   It
would seem that that may be their most natural direction,
because they would thus bisect the right angle made by the
original or vertical stress with ultimate or horizontal strain.
Now, if one diagonal of a square were in tension, the other
would be in compression, because a rectangular or other
figure of four sides, having a tendency to lengthen in the
direction of one diagonal, must also have a tendency to shorten
in the direction of the other, and diagonal stresses in a plate
web, that is to say not conveyed through detached bars, as in
a framed or lattice girder, pervade the whole web, the tensile
passing apparently at right angles to the compressive through
every particle of it.   The resolution of these diagonal stresses
and consequent strains at any given vertical section of the
web would bring us back again to vertical strain.   For,
assuming the diagonal stress, either of tension or compression, as acting upon the particles of the web at an angle of
45° with any vertical section, it is when compared with the
vertical stress on that section as 1 to $\sqrt{2}$, or 1 to 1.4142, when
resolved by the parallelogram of pressures, because the vertical stress due to the load is split into two diagonal stresses,
one of tension and the other of compression, and these in a
solid beam or in a plate web may be regarded as being continuously retransmitted as vertical stress.

Diagonal tensile or compressive stress having been

shown to be less in any one direction in a beam than vertical or shearing stress, in the proportion of 1 to 1.4142, any further notice of diagonal stress and strain becomes unnecessary until the subject of framed and lattice beams is approached.

CHAPTER II.

THE examples given in this chapter are applicable to all
solid beams and webbed single span girders.

Beam with a Central Load and a Support at each End.

23. Vertical Stress. When a beam is horizontally
supported at each end, and loaded at the centre of its span,
the vertical pressure of the load becomes halved at the centre
of the beam, and each half transmitted in an opposite direc-

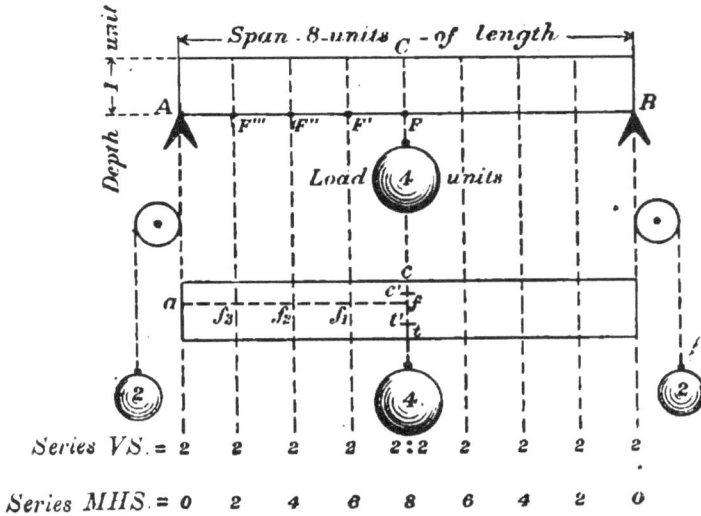

Fig. 9.

tion through the beam to the points of support. Thus as the
centre of gravity of the load is equidistant from the supports

there will be an equal vertical pressure of half the load upon each support. Let A B, Fig. 9, represent a beam of an uniform depth of 1 unit and a length of 8, and assume a load of 4 units at its centre, then with B as the fulcrum of the lever B A by the equality of moments 4 × F B = 2 × A B, therefore there must be a pressure of 2 units at A. In the same way by taking A as the fulcrum of the lever A B there is shown to be a pressure of 2 units at B, and as the stress of one-half the weight of the load passes through the medium of the beam from F to A there must necessarily be at any intermediate vertical section of the beam a vertical or shearing stress of 2 units, Series VS, Fig. 9.

Thus in this example the *effective** vertical stress being 2 at a support, and half the length of the beam being 4 in units of the depth, the *effect* of that stress upon that length = 2 × 4 = 8, or the moment of horizontal stress upon the central section of the beam as stated in the next article, or upon that of each table in the case of a plate girder.

The lower diagram in the figure is appended simply to illustrate the fact that *pressure* and *resistance* are equal and opposite, and therefore convertible terms. The beam is shown suspended at each end by a chain passing over a pulley and carrying a balance weight. Now the tensile stress in each of these chains, balancing and at the same time tending to lift the beam bodily upwards, represents and is precisely the same in amount as the resistance of each support to the pressure upon it, caused by the load upon the beam, as shown in the upper diagram.

It follows that the resistance or vertical reaction of one

---

* With a distributed load the *effective* vertical stress at a support is half that of the actual load upon it (**27**).

support must meet at the central vertical section of the beam precisely the same amount of reaction set up by the other support, and these forces being equal, and acting in the same direction in parallel vertical planes, there will be no shearing strain whatever in the central vertical plane in which they meet, and unite in resisting the direct vertical stress of the load.

24. Horizontal Stresses. As the depth of the beam Fig. 9 = 1 unit, assume A F C to be a bent lever with its fulcrum at F, then the moment of horizontal stress produced on C F by the reaction of 2 at A will be $4 \times 2 = 8$, and if the whole of the stress were at C its amount would also be 8.

If the length of the beam were 4 only the moment of stress would be 4. So that with half the length of beam there is half the horizontal stress on the central section.

The result will be the same if the neutral axis or any other point in the vertical section be taken as the fulcrum. For let any point f in the lower diagram be the fulcrum of a lever of three arms $a f$, $f c$, $f t$, then the stress at any point c, c', t', or t will vary in the direct proportion of the length of the arm $a f$, and therefore in direct proportion to the length of the beam.

In any beam loaded at its centre the amount of horizontal stress at any vertical section varies in direct proportion to its distance from the nearest point of support. For if any number of points, as F′ F″ F‴, are assumed as fulcrums, and vertical lines drawn from these points upwards to represent the vertical arms of a series of bent levers and at the same time vertical sections of the beam, then the strain on each of these sections will be in the direct proportion of its distance from the extreme end A of the horizontal arm of each such lever. Now assume that in the 8-feet beam these sections are taken at 1, 2, and 3 units from A, and having found the moment of stress at the central section $= \dfrac{2 \times 4}{1} = 8$ then the moments at the other sections will relatively $= 2, 4,$ and 6, and in a similar beam 4 feet long with the same load the stress on the four corresponding sections would be represented by the proportionate numbers 1, 2, 3, and 4.

In the same way it may be shown that horizontal stress in the other half of the beam is at each relative vertical equal and opposite, so that if stress in one-half is assumed to be that of action, that in the other half will be stress of reaction. Fig. 9, Series MHS.

It has been shown, therefore, that in any beam loaded at its centre, the horizontal stress caused by the load is greatest at the centre of the beam, that it decreases in terms of arithmetic progression towards each support, there to vanish, and that the strength of a beam of uniform section loaded at its centre is *inversely as its length.*

Beam with a Central Support and a Load at each End.

25. Let the same beam A B, 8 units long and 1 deep, Fig. 9, be supported at its centre F, and loaded with a weight of 2 units at each end as shown Fig. 10. Under these new conditions what were resistances when

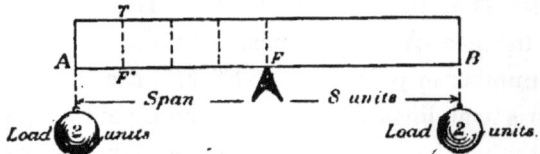

Fig. 10.

the beam was supported at each end have now become loads, and the load itself has become a resistance. Now these terms are convertible because they express equal and opposite forces, as has been shown (**23**). The condition of the beam, therefore, Fig. 10, is nothing more nor less than that shown in Fig. 9, inverted. There is now a vertical resistance of 4 units at F caused by the central support, and it is apparent that the horizontal stresses in the beam under the new are equal to and coincide with those under the previous conditions, with this difference, that their direction is reversed, for the upper part of the beam, which was in compression, is now in tension, and the lower part, which was in tension, is now in compression. The vertical stresses remain also in amount precisely the same as before, but are reversed. For instance, any part A F′ T of the beam has now a tendency to shear off downwards—before, upwards.

It follows that a beam supported at its centre will carry half the load at each end that it is capable of carrying at its centre when supported at the two ends.

Beam with a Distributed Load and a Support at each End.

Let A B, Fig. 11, be a beam 8 units long and 1 deep, supported at each end A and B, and carrying a load of 16 units equally distributed over the entire length.

26. The Vertical Stress occasioned by the pressure of the load is equally divided at the centre of the span, and transmitted by the beam in opposite directions to the two points of support, at each of which it is equal to half the weight of the load. From these points vertical stress in the beam decreases in an arithmetical ratio to nothing at the centre of the span. The effect of this stress arising from a central load having been determined (**23**), we may adopt the same mode of treatment in determining its effect when caused by a distributed load.

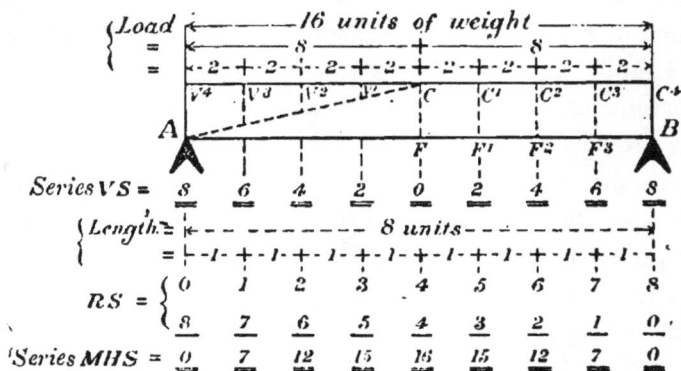

Fig. 11.

Divide the half length of the beam V^4 C into units at points V^3, V^2, V^1; and from the point C at the centre of the beam to the point A draw the diagonal C A. Now as half the load or 8 units is distributed between the points C and V^4, there are two units between C and V^1, and as the vertical stress of these has to be transmitted to the end V^4 A of the beam, it must pass through the beam in that direction, and therefore there is stress of 2 units at V^1. Also between V^1 and V^2, two other units of the load in like manner cause a stress of 2 units at V^2, but the stress of the first two units has also to pass the same point, consequently there will be a vertical stress of 4 units at V^2, and so on to 8 units at the end of the beam. Therefore if the length of the line V^4 A be taken to represent 8 units, or the stress at that perpendicular, the length of any other vertical line drawn from any point in the line V^4 C across the triangle A V^4 C will represent the vertical stress at that part of the beam. Consequently the stress at C, V^1, V^2, V^3, V^4 = 0, 2, 4, 6, 8. It is also evident that stresses of the same amounts exist in the corresponding parts of the other half of the beam, Fig. 11, Series VS.

27. Horizontal Stress. Divide the half C C^4 of the length of the beam into units of length and depth by the dotted lines F C, F^1 C^1, F^2 C^2, F^3 C^3. Now 8 units or the weight of the half of the load extending from C to C^4 may be assumed to act vertically downwards through its centre of gravity, or the line C^2 F^2.

This pressure acting through the bent lever F^2 F C would produce a horizontal tensile stress at C or a compressive stress at F the moment of which about any point in C F would be $8 \times 2 = 16$. But the vertical resistance of 8 at A will cause a horizontal compressive stress at C or tensile stress at F the moment of which will be $8 \times 4 = 32$. The result will be a compressive stress at C or tensile at F with a moment of $32 - 16 = 16$.

The moment of horizontal stress at each of the sections C^1 F^1, C^2 F^2, C^3 F^3 may be similarly determined. At C^1 F^1, there is a stress in one direction caused by the load the moment of which is $6 \times 1\frac{1}{2} = 9$, and one in the opposite direction from the resistance of the point of support with a moment of $8 \times 3 = 24$. The resulting moment of stress is therefore $24 - 9 = 15$. Similarly at C^2 F^2, the moment of stress is $16 - 4 = 12$, and at C^3 F^3 it is $8 - 1 = 7$.

We thus obtain the figures of the Series MHS, 16, 15, 12, 7, and the horizontal stress at the point of support is nil.

At corresponding verticals in the other half of the beam there will exist equal and opposite moments of stress which vanish with the leverage at each end of the beam.   Fig. 11, Series MHS.

Therefore although, as in the case of a beam carrying a central load (**23**), the vertical pressure of the load becomes equally divided at the centre of the beam and transmitted in opposite directions to the points of support and with a load of 16 units either central or distributed, there will be the weight and reaction of 8 units at the support A, the moment of stress upon the central section with the distributed load is 16, or that due to an *effective* weight of 4 only at A. We have therefore the following results :—

1st. The horizontal stress upon the central section caused by a load equally distributed over a beam supported at each end is the same as that which would be caused by one-half of the same load placed at the centre of the same beam.   Therefore if a beam of uniform section will carry a given load at its centre it will carry twice that load equally distributed over its length.

2nd. In any beam of uniform section supported at each end the horizontal stress at the centre caused by its own weight is the same as though half its weight were accumulated at the centre.

3rd. In any beam supported at each end and uniformly loaded, the ratio of the moment of stress upon the central section to that upon any other vertical section is as the square of half the length of the beam to the multiple of the segments into which the beam may be divided.   That this is so may be seen by multiplying the lengths of the segments together at each point in the two Series RS, Fig. 11, the results coinciding with those stated Series MHS.

For as in determining the horizontal stress at the central

vertical of the beam, Fig. 11, we may regard half the weight of the distributed load as statically effective at that vertical, so also the proportion of one-half of the weight of the same load which would be borne at either support if the load were concentrated at any other section may be considered as effective at that section. Because when a load or pressure is assumed to act at unequal distances from the supports, then by the law of the equality of moments (**14**) the fraction of that pressure carried by each support will be inversely as the distance of the pressure from the support, thus—

8 units at $C = 4$ at A and 4 at B then $4 \times 4 = 16$
,, ,, $C^1 = 3$,, A ,, 5 ,, B ,, $3 \times 5 = 15$
,, ,, $C^2 = 2$,, A ,, 6 ,, B ,, $2 \times 6 = 12$
,, ,, $C^3 = 1$,, A ,, 7 ,, B ,, $1 \times 7 = 7$

A similar ratio obtains if we divide the beam into any greater or less number of segments. As an example divide the length of the beam into 16 instead of 8 units, multiply as before, and we have the following Series MHS.

RS. $\begin{cases} 0 & 1 & 2 & 3 & 4 & 5 & 6 & 7 & 8 & 9 & 10 & 11 & 12 & 13 & 14 & 15 & 16 \\ 16 & 15 & 14 & 13 & 12 & 11 & 10 & 9 & 8 & 7 & 6 & 5 & 4 & 3 & 2 & 1 & 0 \end{cases}$

MHS. 0 15 28 39 48 55 60 63 64 63 60 55 48 39 28 15 0

In practice it is frequently sufficient to determine the horizontal stress at the central vertical only. But for large girders the determination of a complete series of stresses may become necessary from the nature of the work, material, or circumstances, while to be on the safe side the higher stress should be assumed to extend down to the lower; thus a moment of stress of 16 at F should be assumed to extend to F^1, 15 at F^1 to extend to F^2, 12 at F^2 to extend to F^3, and 7 at F^3 to extend to the end B of the beam.

Beam with a Distributed Load and a Central Support.

Let the beam A B, 8 units long and 1 deep, Fig. 11, be supported at its centre F, and carry the same load as before of 16 units equally distributed over its entire length as shown, Fig. 12.

28. The Vertical Stress caused by the pressure of the load under this arrangement instead of being nothing at

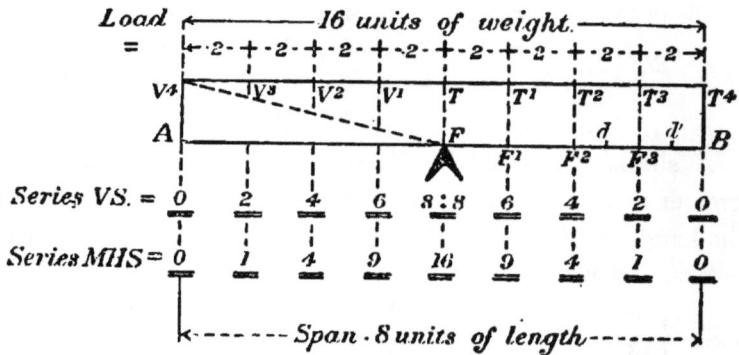

Fig. 12.

the centre of the beam and increasing to half the load at each end of the beam, as in the last example, is now nothing at each end from whence it increases in a direct arithmetical ratio towards the centre of the beam where the stresses caused by the two halves of the load unite in a stress equal to the whole load, and where there is no shearing but compressive stress as the weight of the load is there met by a direct and equal resistance from the point of support.

Halve the length of the beam by the vertical T F, draw the diagonal V^4 F, divide the side V^4 T of the triangle V^4 T F into units of the depth of the beam by points at V^3, V^2, V^1, and draw verticals from those points to the line V^4 F.

Now the weight of the two units of the load extending from V^4 to V^3 has to be transferred by the beam to the support at F, and therefore there must be a stress of two units at V^3. In the same way it will be seen that the stress at V^2 and V^1 is 4 and 6, and at the centre T it will be 8 plus that due to the other half of the load, or $8 + 8 = 16$ units.

If therefore the length of the line T F be taken to represent 8, the length of the other verticals, as for instance V^1, V^2, V^3, drawn across the triangle V^4 T F, will represent the vertical stress in the beam at those lines, and the same stresses will exist in the corresponding parts of the other half of the beam. Fig. 12, Series VS.

29. Horizontal Stress. Divide the half of the length T T^4 of the beam into units of its depth, by the dotted lines T^1 F^1, T^2 F^2, T^3 F^3, and halve the spaces F^2 F^3 and F^3 B in points d and d'.

1. The moment of stress at the central section T F is that occasioned by half the load or 8 units acting vertically through its centre of gravity and the line T^2 F^2, as upon the end F^3 of a bent lever F^2 F T, the fulcrum being at F ; and therefore $8 \times 2 = 16$.

The stress on the central section is therefore directly as the length of the arm F F^2 of the lever, and consequently directly as the length of the beam, as in the example already given **(27)**.

2. The moment of stress at the section T^1 F^1 is that due to $\frac{3}{4}$th of the load, being the portion from T^1 to T^4, or 6 units, acting vertically through its centre of gravity upon the end

d of the bent lever d F^1 T^1, the fulcrum being at F^1; or $6 \times 1\frac{1}{2} = 9$.

3. The moment of stress at the section T^2 F^2 is similarly that due to the action of $\frac{1}{4}$th of the load, or 4 units, being the portion from T^2 to T^4, at a distance of 1 unit, and is therefore 4.

4. In the same manner the moment of stress at the section T^3 F^3 from 2 acting at d' is $2 \times \frac{1}{2} = 1$.

The moments of horizontal stress thus determined are therefore 16, 9, 4, 1, or in direct proportion to the squares of the distances from the end B of the beam, the distances being 4, 3, 2, 1. In like manner the stresses in the other half of the beam are equal and opposite. Figure 12, Series MHS.

It has therefore been shown :—

1st. That in a beam supported at its centre and uniformly loaded throughout its length, *horizontal stress* is greatest at the centre where it varies *directly as the length of the beam.* The *strength of the beam* is therefore *inversely as its length.*

2nd. That horizontal stress is the same in amount at the centre of the beam as if the beam were supported at its ends and similarly loaded, but that it decreases towards each end of the beam in the ratio of the square of the distances from the end, and *not* as the multiple of the segments as in the former case.

Consequently a beam of *uniform section,* whether supported at its centre, or at each end, will carry the same load uniformly distributed, or twice the weight that it would be able to carry at its centre when supported at the two ends. But in the second condition the stresses right and left of the centre are greater.

Beam with a Non-central Load and a Support at each End.

Let A B, Fig. 13, be a beam of uniform section 10 units long and 1 deep, supported at each end A and B, and carrying a load of 20 units at a point C, 7 from A and 3 from B.

30. The Vertical Pressure of the load travels in unequal fractions from the point C to each support in the inverse proportion of the distance of the support from C, the vertical of the centre of pressure of the load. Thus as

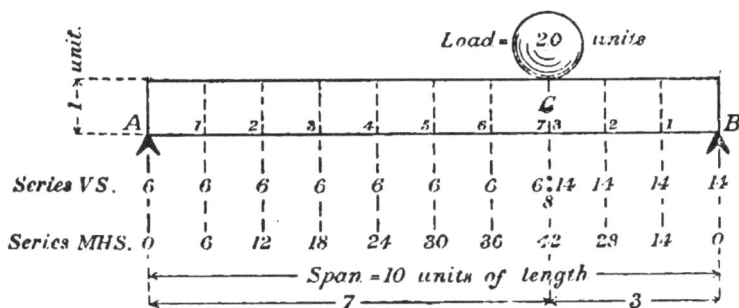

Fig. 13.

the load is 20 and the span 10, and as A is 7 distant from C, there will be a stress of 14 at B, and as B is 3 distant from C, there will be a stress of 6 at A. For let A B be a lever with the fulcrum at B, then by the equality of moments the resistance at A $= \dfrac{20 \times 3}{10} = 6$, and in the same way the resistance at B $= \dfrac{20 \times 7}{10} = 14$. The vertical stress caused by the load therefore becomes unequally divided in the inverse ratio of the segments into which the beam is divided by the centre of pressure, thus, assuming the load to be 1,

$\frac{6}{20}$ = .3 passes from C through the beam to the support at A, and $\frac{14}{20}$ = .7 passes from C in the same way to the support at B. Fig. 13, Series VS.

Now the vertical shearing stress caused by the resistance at A = 6 extends from A to C, and the vertical shearing stress caused by the resistance at B = 14 extends from B to C. Two unequal shearing stresses therefore meet at C, so that the true shearing stress in the theoretical vertical plane which at C divides the beam into two segments is equal to the difference of the two, or = 14 − 6 = 8.

31. The Horizontal Stress at any cross section of the beam may readily be determined in proportional amounts, as shown in Article **24.** Thus, the beam having been longitudinally divided into units, commencing with the vertical stress of 6 at A, 6 × 0 = 0, next 6 × 1 = 6, and so on to the centre C, where 6 × 7 = 42. In the same way, commencing with the vertical stress of 14 at B, 14 × 0 = 0, next 14 × 1 = 14, then 14 × 2 = 28, while at the centre it is 14 × 3 = 42. The moment of stress of action and reaction at the section C has therefore been shown to be equal and opposite, and the same must necessarily be the case at all corresponding sections right and left of C. Fig. 13, Series MHS.

It will be seen, therefore, that the horizontal stress at C, the centre of pressure of the load, is as the multiple of the segments into which that point may divide the beam, and that consequently it will be the greatest when the load is central between the supports, while in all cases it will diminish from the centre of pressure of the load in terms of arithmetic progression from a maximum at C to nothing at each support.

Corollary. It follows conversely that if the same beam were supported at C and loaded with 6 units at the end A and with 14 at the end B, Fig. 14, equilibrium would be maintained, and that the

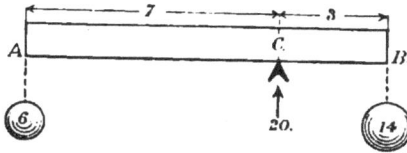

Fig. 14.

amounts and ratios of the stresses would be the same as before, but that the relative conditions of tension and compression would be reversed (**25**).

Beam with two Equidistant Loads, and a Support at each End.

Let A B, Fig. 15, be a beam of uniform section 10 units long and 1 deep supported at each end A and B, and carrying

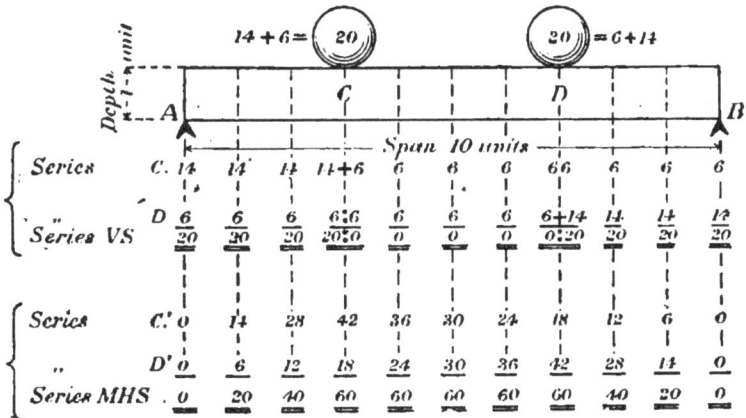

Fig. 15.

a load of 20 at C, 3 distant from A, and a load of 20 at D, 3 distant from B.

32. Vertical Stress. The manner in which the direct stress of a non-central load is divided having been shown (**30**), place the stresses due to the weight of the load at C in a Series C, Fig. 15, and those due to the weight of the load at D in a corresponding Series D. Now at any section between the verticals C and D the stress thus given is in each case 6, but when these stresses are taken in conjunction, it will be seen that there is no shearing stress at all in that part of the beam, because the two are equal and act in the same direction, and there is no vertical resistance between those points, but the load of 20 at C meets with a like resistance at A, and that at D with a like resistance at B. Therefore the two series of assumed stresses occurring between sections C and D of the beam may be subtracted the one from the other, and, as already said, the shearing stress in that part of the span is nil, while that set up by the load of 20 at C extends to A, and that due to the load of 20 at D extends to B. Fig. 15, Series VS.

33. Horizontal Stresses. The mode of determining these for either the load at C or D having been given (**31**), it only remains to place the moments due to the load acting at C in a Series C′, Fig. 15, and those due to the load acting at D in a corresponding Series D′, and then to add the amounts together, placing in a final series the aggregate of the united stresses due at each section to the two loads. Thus Fig. 15, Series MHS., gives the amounts of the combined moments of stress due to the two loads at each unit of length of the beam. It will be observed that from centre to centre of the two loads or between the verticals C and D the moment of stress remains constant, because, as already stated, there exists no vertical resistance between those sections.

D

Corollary. It follows conversely that if the same beam were supported at C and D, and loaded with 20 units at each end A and B, Fig. 16, the amounts and ratios of all stresses would remain the same as before, but that the relative conditions of tension and compression would be reversed (**25**). It will also be apparent that there cannot be any vertical or shearing stress between C and D. Between these verticals there is theoretically no need of any web.

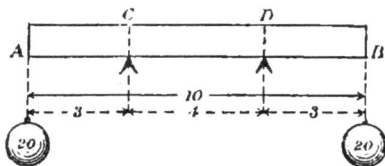

Fig. 16.

In the case of a cast-iron girder, however, it would be essential from the nature of the material and its section to continue the web between C and D, and the same would also be necessary in a plate girder in order to stiffen vertically that table which has to resist compressive stress.

Beam carrying Loads at various Points and supported at each End.

Let A B, Fig. 17, be a beam 12 units long and 1 deep, supported at each end, A and B, and carrying six various loads at the verticals C, D, E, F, G, and H, placed 1, 4, 6, 7, 9, and 10 units from A, and = 12, 3, 2, 12, 4, and 6 units in weight respectively.

34.   Vertical Stresses.   Referring to **32**, it will be evident that these may be readily determined if we proceed first to ascertain the total resistance each support has to present to the pressure caused by the six loads.

Thus (**30**) 11 units of weight (Fig. 17, * Series *vs*) are transferred from C to A + 2 from D, + 1 from E, + 5 from F, + 1 from G, + 1 from H = a total of 21 units of weight at A. Again 1 is transferred from C to B, + 1 from

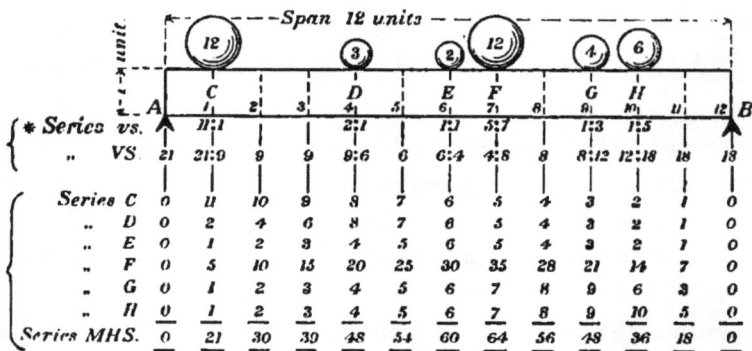

Fig. 17.

D, + 1 from E, + 7 from F, + 3 from G, + 5 from H = a total of 18 units of weight at B as given Series VS. Now having ascertained that there is a stress of 21 at A, deduct in succession the load at C, D, and E ; then the stress from A to C = 21 − 0 = 21, from C to D = 21 − 12 = 9, and from D to E = 9 − 3 = 6. The aggregate thus deducted = 12 + 3 + 2 = 17, but there is a weight of 21 on the support A, so that 4 units pass from the load of 12 at F to the support A, and the stress from E to F = 4, leaving at F 12 − 4 = 8. In the same way having a stress of 18 at B, deduct in succession the loads at H and G, then the stress from B to H = 18 − 0 = 18, from H to G = 18 − 6 = 12, and from G to F = 12 − 4 = 8. The aggregate thus deducted is 6 + 4 = 10, but as there is a weight of 18 on the support B, the 8 units we have left at F must pass

to the support B, and as already found, the stress from G to
F = 8. Fig. 17, Series VS.

As the units of vertical stress part company right and left
in the unequal fractions $\frac{4}{12}$ and $\frac{8}{12}$ at the section F, the vertical shearing stress at that section = 8 − 4 = 4 (**30**).

35. The Horizontal Stresses caused by each
load are also readily determined as shown (**31**). Place
in relative series C, D, E, F, G, and H, the moment of
stress so given at each unit of length of the beam. Add
up and place the amounts in a final Series MHS, Fig. 17,
which thus gives at each unit of length of the beam the
moment of horizontal stress caused by the six loads.

Beam with a Concentrated Rolling Load.

Let A B, Fig. 18, be a beam 10 units long and one deep,
supported at each end A and B, and carrying a concentrated
rolling load of 10 units.

36. Vertical Stress. It has been shown (**23**)
that when the centre of gravity of a concentrated load coincides with the central vertical of the beam, the vertical stress
due to the weight of the load becomes equally divided, and
each half transmitted in an opposite direction to the points of
support. Further (**30**), that when the centre of gravity of
the load coincides with any vertical section of the beam
other than the central one, vertical stress so caused becomes
unequally divided and transmitted from that section to each
support in unequal fractions in the inverse proportion of the
distance of the section from the support.

Now bearing in mind these premises, let the load of 10

units at the end A of the beam roll to the other end B. Then in successively passing the verticals 1, 2, 3, 4, 5, 6, 7, 8, 9, its effect at each will be the fractional vertical stresses given in Fig. 18, Series VS.

It will further be seen by reading from either support towards the other, that each of these fractional stresses increases in a ratio of direct arithmetical progression until it becomes equal to the whole load at each support. Thus starting from both ends, these fractional stresses are respectively 0, 1, 2, 3, 4, and 5 at the centre of the beam, and continuing past the

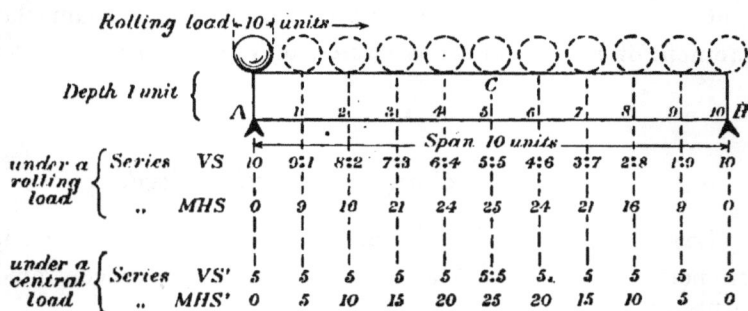

Fig. 18.

centre in each direction, 6, 7, 8, 9, and 10 respectively. Whereas if the same load were to remain stationary at the central vertical C, vertical stress would be represented by 5 throughout, Fig. 18, Series VS'. Comparing Series VS with Series VS', it will be noted that the difference implies considerable theoretic modification in the web of a girder in order that it should properly resist vertical stress caused by a concentrated rolling load.

37. Horizontal Stress. It has been shown **(31)** that in the case of a concentrated load carried by a beam at any point between the two supports, the moment of stress is directly as the multiple of the segments into which that point divides the beam. Now a concentrated rolling load, in moving from end to end of a girder, must pass every assumed vertical. If, therefore, we multiply the segments into each other, at each such vertical in succession, the moment so determined for each will proportionately represent the stress caused by the load in passing, Fig. 18, Series MHS. Comparing the moments thus determined with those in the Series MHS′ of the moments of horizontal stress at the same points when the load is stationary at the central vertical C, it will be seen that in the former case the moments are, as in the case of a distributed load, represented by the multiple of the segments giving the ordinates of a para-bola, whereas in the latter the moments increase in a direct arithmetical ratio from each support to the centre of the beam. Therefore the strength of the *tables* of a girder intended to carry a concentrated rolling load, passing longitudinally over it, should between its centre and supports be greater than that required for an equal central stationary load in the relative proportions of the resolved moments as given in Figure 18.

Beam with Two Rolling Loads coupled.

Let A B, Fig. 19, be a beam 10 units long and one deep supported at each end, A and B, and carrying two rolling loads of 10 units each, coupled together at a distance of 2 units from centre to centre of each.

38. **The Vertical Stresses** will in this instance become incident at the assumed respective verticals simultancously in pairs as bracketed a, b, c, d, e, and f in the figure. Now all that is necessary for the determination of these stresses is to find (**30**) the amounts of the fractions of

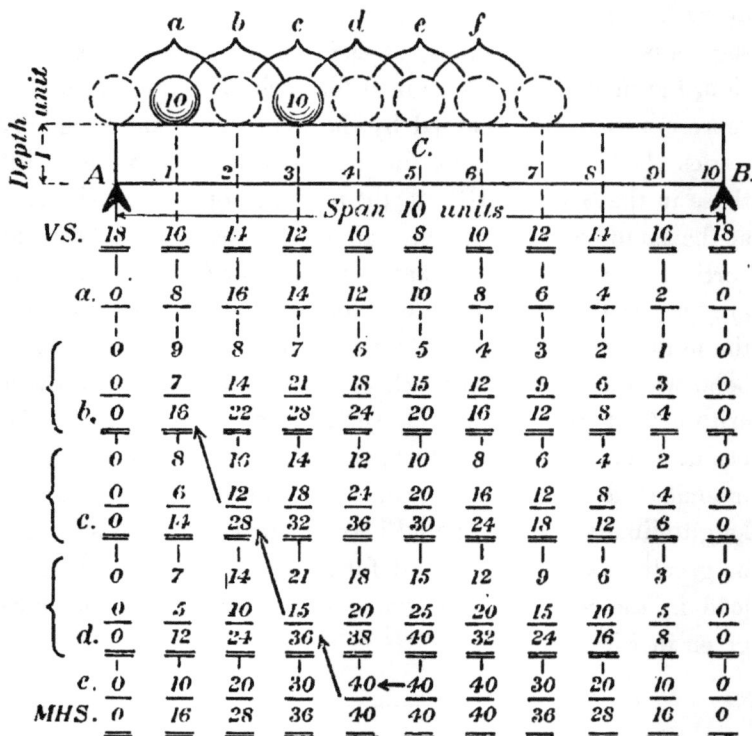

```
            a     b     c     d     e     f

Depth
 1 unit                            C.
      A   1|  2|  3|  4|  5|  6|  7|  8|  9|  10| B.
                    Span 10 units

VS.   18  16  14  12  10   8  10  12  14  16  18

a.     0   8  16  14  12  10   8   6   4   2   0
       0   9   8   7   6   5   4   3   2   1   0

b.     0   7  14  21  18  15  12   9   6   3   0
       0  16  22  28  24  20  16  12   8   4   0
       0   8  16  14  12  10   8   6   4   2   0

c.     0   6  12  18  24  20  16  12   8   4   0
       0  14  28  32  36  30  24  18  12   6   0
       0   7  14  21  18  15  12   9   6   3   0

d.     0   5  10  15  20  25  20  15  10   5   0
       0  12  24  36  38  40  32  24  16   8   0
       0  10  20  30  40  40  40  30  20  10   0

e.     0  10  20  30  40←40  40  30  20  10   0
MHS.   0  16  28  36  40  40  40  36  28  16   0
```

Fig. 19.

stress which travel in the direction of one of the supports, say the support A.

Assume the pair of loads to be moving from A towards B, and to be in the first positions bracketed a, then for the stress at the end A, to the load 10 at A add 8, or the fraction due from the load at vertical 2, and $10 + 8 = 18$. Next when the loads are in the second position or insistent at verticals 1 and 3, the stress at vertical 1 is·9 + 7 = 16 ; with the loads at 2 and 4, that at 2 is 8 + 6 = 14 ; with the loads at 3 and 5, that at 3 is 7 + 5 = 12 ; with the loads at 4 and 6, that at 4 is 6 + 4 = 10 ; and with the loads at 5 and 7, that at 5, or the central vertical C of the beam, is 5 + 3 = 8. Thus, Fig. 19, Series VS, completed from C to B in reverse order, gives the vertical stress due at each assumed vertical as the coupled loads pass from end to end of the beam.

39. Horizontal Stresses. Assume as before the coupled loads to be moving from A towards B. Then at each assumed vertical resolve the moments of stress (**31**) due to each of the loads in passing the positions indicated by the letters *a, b, c, d, e, add the amounts* due to the coupled pairs together, and place them in Series *b, c,* and *d* (Fig. 19). It will be observed that Series *a* does not fall within a coupled series, nor within the final result ; and that Series *e* can be obtained by one operation (**33**). Now commencing from the centre and then following the results in Series *e, d, c, b,* in the order just stated and in the direction indicated by the arrows, we have the maximum stress to which each assumed vertical throughout the half length of the beam has been subjected during the passage of the combined loads from A until their centre of gravity coincided with vertical 5, or the centre C of the span. Write these down for the first half of the span, and repeat the same for the second half in reverse order, and we obtain Series MHS, or the moments of maximum horizontal stress at each assumed vertical caused by the coupled load in passing from end to end of the beam.

In this manner the stresses in a girder caused by the fore

and aft wheels of a loaded trolley passing over it longitudinally may be determined, whether the loads carried by the wheels be equal or unequal.

Beam with an uniformly Distributed Load advancing upon it.

Let A B, Fig. 20, be a beam 8 units long and 1 unit deep supported at each end A and B, and subjected to a continuous load of 64 units or eight to each single unit of the length.

40. Vertical Stresses. Divide the beam into units of length by verticals 1, 2, 3, 4, 5, 6, 7, and assume the continuous load to advance upon the beam unit by unit in length from either end, say from the end A.

Assume 8 load units or a weight = 8 in respect of each unit of length to act vertically downwards through its centre of gravity. Then proceed to determine (**30**) the vertical stresses caused by the load carried by each successive unit of length. Add together at each vertical the results thus obtained *until* the fore end of the advancing load coincides with the central vertical 4 at C. Beyond that vertical the stress caused by any further advance must be subtracted at any point between that of advance and the corresponding point already passed on the left side of the centre of the span (**32**).

The Series VS 1, 2, 3, 4, 5, 6, 7, 8 will then give the stress at each vertical, due to each successive fractional advance of the load, until there are 8 load units upon each unit of length. Commencing from the support A, the bracketed figures indicate the position of the leading fraction of the load for each series. Between the centre of gravity of

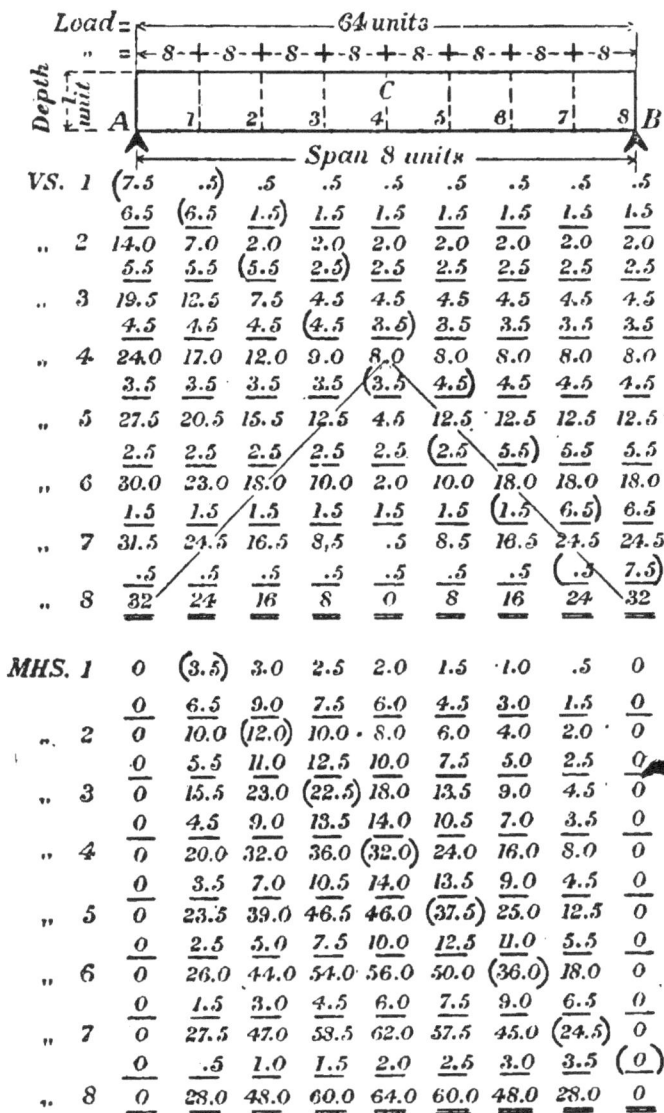

Load = 64 units
" = 8 + 8 + 8 + 8 + 8 + 8 + 8 + 8
Depth 1 unit

A | 1 | 2 | 3 | 4(C) | 5 | 6 | 7 | 8 B

Span 8 units

VS.	A	1	2	3	4	5	6	7	8/B
1	(7.5)	.5	.5	.5	.5	.5	.5	.5	.5
	6.5	(6.5)	1.5	1.5	1.5	1.5	1.5	1.5	1.5
2	14.0	7.0	2.0	2.0	2.0	2.0	2.0	2.0	2.0
	5.5	5.5	(5.5)	2.5	2.5	2.5	2.5	2.5	2.5
3	19.5	12.5	7.5	4.5	4.5	4.5	4.5	4.5	4.5
	4.5	4.5	4.5	(4.5)	3.5	3.5	3.5	3.5	3.5
4	24.0	17.0	12.0	9.0	8.0	8.0	8.0	8.0	8.0
	3.5	3.5	3.5	3.5	(3.5	4.5)	4.5	4.5	4.5
5	27.5	20.5	15.5	12.5	4.5	12.5	12.5	12.5	12.5
	2.5	2.5	2.5	2.5	2.5	(2.5	5.5)	5.5	5.5
6	30.0	23.0	18.0	10.0	2.0	10.0	18.0	18.0	18.0
	1.5	1.5	1.5	1.5	1.5	1.5	(1.5	6.5)	6.5
7	31.5	24.5	16.5	8.5	.5	8.5	16.5	24.5	24.5
	.5	.5	.5	.5	.5	.5	.5	(.5	7.5)
8	32	24	16	8	0	8	16	24	32

MHS.	A	1	2	3	4	5	6	7	8/B
1	0	(3.5)	3.0	2.5	2.0	1.5	1.0	.5	0
	0	6.5	9.0	7.5	6.0	4.5	3.0	1.5	0
2	0	10.0	(12.0)	10.0	8.0	6.0	4.0	2.0	0
	0	5.5	11.0	12.5	10.0	7.5	5.0	2.5	0
3	0	15.5	23.0	(22.5)	18.0	13.5	9.0	4.5	0
	0	4.5	9.0	13.5	14.0	10.5	7.0	3.5	0
4	0	20.0	32.0	36.0	(32.0)	24.0	16.0	8.0	0
	0	3.5	7.0	10.5	14.0	13.5	9.0	4.5	0
5	0	23.5	39.0	46.5	46.0	(37.5)	25.0	12.5	0
	0	2.5	5.0	7.5	10.0	12.5	11.0	5.5	0
6	0	26.0	44.0	54.0	56.0	50.0	(36.0)	18.0	0
	0	1.5	3.0	4.5	6.0	7.5	9.0	6.5	0
7	0	27.5	47.0	53.5	62.0	57.5	45.0	(24.5)	0
	0	.5	1.0	1.5	2.0	2.5	3.0	3.5	(0)
8	0	28.0	48.0	60.0	64.0	60.0	48.0	28.0	0

Fig. 20.

this fraction and the support B the stress necessarily remains constant in each series.

Series VS, 8 gives the stresses when the load has advanced so as to cover the whole span, and which it is to be noted are those due to an evenly distributed load of 64 units.

The greatest stress at the central vertical occurs when the load has just advanced up to it. Then at that vertical it is one quarter of the ultimate maximum stress at an abutment. Whereas with an equally distributed load covering the whole span the vertical stress at the centre of the span is nil.

For let the weight of a load uniformly distributed over the whole span be 2 w, then the weight on each abutment would be w. Now the weight of the same load extending from an abutment over one-half of the same span would also be w with its centre of gravity at a distance of one-fourth of the span from the abutment. Therefore the weight on the further support and consequently the vertical stress at the centre of the span would be $\frac{w}{4}$ or in the example given $\frac{32}{4} = 8$, Series VS 4.

The maximum stress which occurs at each given vertical is indicated in Series VS, Fig. 20, by two diagonal lines.

From a line A B, Fig. 21, representing the half-span, set up at equal distances in relative order to an increased scale as ordinates these maximum stresses and through the extreme points thus found draw the curved line C D.

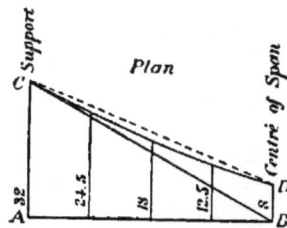

Fig. 21.

Then as C A is the stress at an abutment, and D B that at the centre of the span, following the curved line C D the figure C D B A proportionately represents the theoretic horizontal sectional *web area* required for an equally distributed advancing load, and in the same way the triangle C B A proportionately represents the horizontal sectional *web area* required by the same load when extending over the whole span. The first of these areas is $\frac{1}{6}$ greater than the second.

41. Horizontal Stresses. Having already determined as given in Series VS the vertical stress at each abutment due to each unit by unit advance of the load, commence from the support A, Fig. 20, and determine in succession (**31**) the moment of horizontal stress at each given vertical as the leading end of the load becomes incident with it.

Thus when the load has advanced to vertical 1, the vertical stress at B is .5, and this multiplied by leverage in units of length gives Series MHS 1. Again, when the load has advanced from 1 to vertical 2, that portion of it will cause a vertical stress of 1.5 at B, and 6.5 at A, and each of these multiplied by their respective leverages gives the next or an intermediate series, which added to Series 1 gives Series 2, or the stress at each unit of length when the load covers the portion of the beam extending from A to vertical 2.

In the same way determine in succession the moment of stress at each given vertical as the leading end of the advancing load becomes incident with it until the whole beam is covered.

We shall then have obtained Series MHS 1, 2, 3, 4, 5, 6, 7, 8.

The bracketed figures indicate in each series the termination of the load's advance, from which point the stress diminishes in the unloaded segment of the beam in terms of arithmetic progression to nothing at the point of support.

When the advance end of the load has arrived at vertical 8, the whole beam is covered with an evenly distributed load of 8 units to one of length, and the moments of stress are then directly as the multiple of the given segments (**27**). For instance, by dividing the length of the beam into 16 units, and then multiplying each pair of segments together, we should thus obtain the same results as those stated in Series MHS 8.

It will be noted that the maximum horizontal stress at each vertical occurs when the load completely covers the beam, and that this has been shown in the last Article to be otherwise with vertical stress.

CANTILEVERS.

Suppose a beam instead of being supported at its centre to have one end built into a wall up to that point, then the remaining or projecting portion will be a cantilever (**5**).

The projecting parts of cantilevers fall, therefore, under precisely the same static conditions of stress and strain, as either half of a beam supported at its centre. It is, however, desirable to note some collateral particulars attending the use of cantilevers.

42. As an example, let A B, Fig. 22, be a cantilever $1\frac{1}{2}$ units deep projecting 4 units from a wall at B into which it is continued for **2** units, and carrying *a load* of 9 units *at its outer end* A. Assume the resistance

presented to the turning of the beam round a point at D
by the superincumbent weight
of the wall to be at C, then
a vertical or shearing stress of
the load $= 9$ will be constant
from A to B (**21, 25**), and as
A B $= 4$ and B C $= 2$ there will
be an additional stress at D due to
the resistance at C, $\dfrac{9 \times 4}{2} = 18$, but

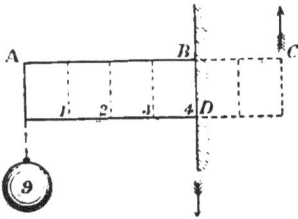

Fig. 22.

$18 - 9 = 9$ (**30**) will be still the shearing stress at vertical 4,
while at the same time $18 + 9 = 27$ is the total vertical
pressure at D, tending to thrust down the wall in the direction of the arrow. It consequently follows that as the resistance at C has been shown to $= 18$, there is an upward
pressure of 18 at C as indicated by an arrow, tending to
overturn the upper part of the wall about the point D.

The moments of horizontal stress caused by the load
are (**24, 25**) relatively at verticals 1, 2, 3, 4, $\dfrac{9 \times 1}{1.5} = 6$,
$\dfrac{9 \times 2}{1.5} = 12$, $\dfrac{9 \times 3}{1.5} = 18$, $\dfrac{9 \times 4}{1.5} = 24$, and the moment of reaction upon vertical 4 caused by the resistance of 18 at C $=$
$\dfrac{18 \times 2}{1.5} = 24$.

43.   Assume the same cantilever, Fig. 22, to carry a
load equally distributed from A to B. It will then with the
same horizontal stress at vertical 4, which coincides with the
assumed point of support, carry twice as much as when
loaded at the end, or a load of $9 \times 2 = 18$ (**29**).

As there are $\dfrac{18}{4} = 4\frac{1}{2}$ units of load to 1 of length, the
shearing stress at verticals 1, 2, 3, 4 will relatively be $4\frac{1}{2}$,
9, $13\frac{1}{2}$, 18 (**28**). Now the mean leverage of a distributed

load measured from any given point is the distance of its centre of gravity from that point (**27**), therefore in this case the centre of gravity of the load being at vertical 2, and the fulcrum at D, it follows that the vertical resistance required at C $= \dfrac{18 \times 2}{2} = 18$, or the same as with a load of 9 at the end A of the cantilever. The total vertical pressure at D tending to thrust the wall downwards in this case is therefore $18 + 18 = 36$, and the upward pressure at C tending to upset the upper part of the wall about point D will as before be 18.

The moment of horizontal stress caused by the distributed load is (**29**) as the square of the distance from the end A, therefore 4^2 or 16 would represent this stress at vertical 4, but as the depth of the cantilever is $1\frac{1}{2}$, the actual stress will be $\dfrac{18 \times 2}{1.5} = 24$ or, $16 \times 1\frac{1}{2}$ at vertical 4, and therefore for the whole series of verticals 1, 2, 3, 4, they will relatively be $1\frac{1}{2}$, 6, $13\frac{1}{2}$, 24.

It follows therefore that given a beam of uniform section of any length carrying a maximum central load, one-half of its length will as a cantilever carry at its outer end half that load, and if the cantilever were of the same length as the original beam, it would then carry one quarter of the original load at its outer end.

In the same way, one-half of a beam will as a cantilever carry one-half of the distributed load carried by the whole beam when supported at each end, or if the cantilever were as long as the whole beam it would then carry one-quarter of that load.

CHAPTER III

44. Beam of Uniform Section placed upon Two Supports, and carrying an equal load at each end, and a load at its centre equal in weight to the two end loads. Determination of the most efficient positions for the supports. Let $a\,c\,b$, Fig. 23, be a platform carrying an equally distributed load of 12 units or 6 on each side of the centre, and let this be

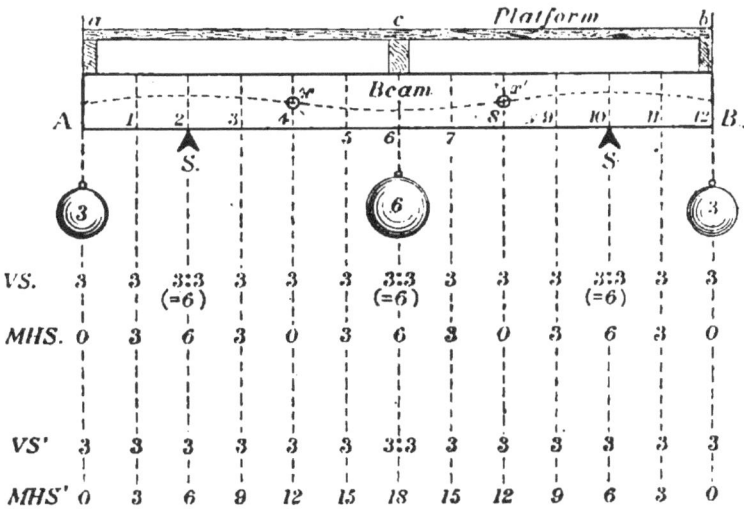

Fig. 23.

assumed to occasion a pressure of 3 units at each of the extremities *a* and *b*, and of 6 units at the centre *c*.*

Let the bearers *a* and *b* be carried at the ends, and the bearer *c* at the centre of a beam A B.   Then the beam A B will carry a load of 3 at each end, and 6 at its centre.   Now the supports S, S of this beam would not be in the most efficient positions if placed at its ends, for although the beam would thus be relieved of all horizontal stress that might be caused by the loads there incident, that occasioned by the central load would be a maximum, and any advance of the supports towards the centre would not only diminish the span between them, but also convert the outer or projecting ends of the beam into cantilevers which it will be seen would further diminish the central horizontal stress.

If the supports were so advanced the beam would tend to be deflected in the manner shown by the dotted line in the figure, the upper part of the beam over each of the supports S, S and the lower part under the central bearer of the platform being in a state of horizontal tension.

The points at which this change of strain takes place are termed points of contrary flexure, *x x′*, Fig. 23, the curvature which is upward on the one side of these becoming downward on the other.   The horizontal stress decreases as they are approached, vanishing at them, and recommencing from nothing, again increases on the other side.

It is obvious that the further the supports S, S are advanced from A and B, the greater will be the horizontal tension over them, and the less that at the centre of the beam.

---

* This will not be absolutely true unless the platform is divided at *c*, but under ordinary circumstances the effect of the rigidity of the platform may be disregarded.

E

The most efficient positions for the supports will therefore be those which will make the horizontal stress at each support exactly equal to that at the centre of the beam.

Now the load of 3 units at each end of the beam will cause a vertical pressure of 3 upon each support S, and the load of 6 at the centre of the beam will also cause a vertical pressure of 3 at each.

There will therefore be a vertical pressure of 6 upon each support, and an equal stress at the centre of the beam. But the pressure upon and the resistance of the supports are equal and opposite, and the latter may be regarded as a load acting upwards instead of downwards.

The points of contrary flexure x x' may therefore be considered as dividing the beam into three spans, A x, x x', and x' B, the central span being supported at the points of contrary flexure, and each of the two outer spans by one of the supports S, S. In order, therefore, that there may be the same horizontal stress at the centre of each of these three spans, the loads at their centres being equal, the lengths of the spans should also be equal, and the supports should be placed at the centres of the two outer spans.

Thus if the length of the beam when divided by units of its depth be 12, each span would be 4 units long, the points of contrary flexure will be at the verticals 4 and 8, and the most efficient position of the supports S, S at the verticals 2 and 10.

The vertical stresses are given in Series VS, and the moments of horizontal stress in Series MHS. Fig. 23.

45. Weight of the Beam, Fig. 23, as a Plate Girder, with End Supports, as compared with

that with Intermediate Supports. The Series VS′
and MHS′ have been appended to the diagram to show the
amounts of vertical and horizontal stress which would exist
if the beam were supported at each end instead of at S, S,
and carried as before 6 load units at its centre.

Assume the resistance of the material per unit of sectional
area to be the same for all stresses. Then, as the amount of
material required per unit of length would vary as the stress,
add together Series VS′ and MHS′. Deal in the same way
with Series VS and MHS, and the totals will give the relative
amount of material required for the two systems of support,
which will be to each other as 144 to 72 or 2 to 1.

Theoretically, therefore, two equally efficient girders when
supported as shown, Fig. 23, could be made with the amount
of material required for one girder of the same total length
if supported at each end.

Practically, however, the result for the intermediate
supports would not be quite so favourable because for the
sake of lateral rigidity it would not be desirable to reduce the
tables of a girder to nothing at the points of contrary flexure.

Now as small girders are usually made of uniform thickness throughout, and *rolled joists* necessarily so, it is for these
only necessary to compare the central transverse stress due to
each of the two systems of support because this the maximum
will determine the weight of the beam per length unit. Thus
the value at the central section, of VS + MHS = 3 + 6 = 9
and VS′ + MHS′ = 3 + 18 = 21 and $\frac{21}{9} = \frac{7}{3}$. Therefore
7 girders supported as shown, Fig. 23, could practically be made
under these conditions with the amount of material required
for 3 girders each of the same total length and of the same
efficiency, but supported at each end.

46. To determine, as before, the most efficient position for the two supports, but under other conditions of loading. Let A B, Fig. 24, be a beam

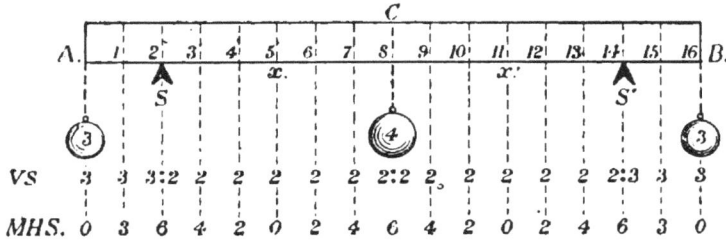

Fig. 24.

of uniform section 1 unit deep carrying 3 load units at each end A and B, and 4 at its centre C. Now the two points of support S S' will necessarily be equidistant from C, and the central load of 4 will cause a vertical stress of 2 at each support, and consequently also the same stress at each vertical of contrary flexure x and x' (Series VS). But (**44**) A x may be treated as a separate beam placed on an intermediate support, and loaded in the present example with 3 units at A, and 2 at x. Therefore $3 + 2 = 5$ is the pressure upon the support S. Make A $x = 5$ units of length, and by the equality of moments the distance of S from A and x will be inversely as the loading at those two extremities, and therefore A S = 2 and S x = 3. The moment of horizontal stress in the beam at the support S will then be $3 \times 2 = 6$. Moreover x x' when treated as a separate beam, with a vertical stress of 2 at x and at x', it follows, in order that there shall be a horizontal stress of 6 at vertical C, that x C should $= x$ S. Therefore in this instance x C = 3, and $3 \times 2 = 6$, the horizontal

stress at C and also that at S. And the half length of the
beam $= 2 + 3 + 3 = 8$ units.

Therefore, given a beam of uniform section loaded in this
ratio, if we divide its length into 16 units the most efficient
position for the two supports will be at 2 units from each
end.

Series MHS gives the moments of horizontal stress from
end to end of the beam.

In the same manner the most efficient positions of the
supports, when assumed to be equidistant from a given
intermediate load, may be readily determined for any given
loads at the ends of the beam.

47. Beam of Uniform Section with Two Supports, and a Distributed Load. To find the most efficient positions for the supports.
Let
A B, Fig. 25, be a beam carrying an equally distributed load
of 2 units to 1 of length over its entire length, and as it has
been shown (**44, 46**) that the most efficient positions of
the supports S S′ cannot be at each end of the beam, let the
cantilever ends A S and S′ B be each one unit in length.

Assume for the present
that the two ends A S and
S′ B alone are loaded, and
that there is no load between
S and S′. Then the moment

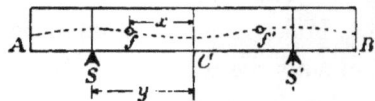

Fig. 25.

of horizontal stress over the supports S and S′ will be
$2 \times .5 = 1$ (**43**), and this stress will be continued between
those verticals (**33,** Cor.), placing the upper part of the
beam in a state of tension and the lower part in a state of
compression.

But when the loading is continued between the points of support S, S', the beam will become deflected as indicated by the dotted line in the figure, and the horizontal stresses will be reversed in the central part of the beam, where the upper part will now be in compression and the lower in tension. It consequently follows that in order to *neutralise* and *reverse* the stress = 1 as first assumed to have been set up in that part of the beam an opposing moment = 2 will be required.

Let y equal the distance between the support S and the centre C of the beam,

Then $2y$ = the load between S and C,

and $\frac{y}{2}$ = the leverage (**27**).

$$\therefore \ 2y\,\frac{y}{2} = y^2$$

let $y^2 = 2$ or the reversing moment required
then $y = \sqrt{2} = 1.4142.$ (1.)

Now as A S = 1, so f C will also be 1 (**44**); therefore the distance of each point of contrary flexure f or f' from the nearest point of support = 1.4142 − 1 = .4142, and the total length of the beam is 4.8284.

With a given length of beam, say for instance 2 units, the relative distance of the points of contrary flexure from the nearest support will be $\dfrac{.4142 \times 2}{4.8284} = .1716.$

An *algebraic solution* of this problem is given, Note (*a*), p. 57.

48. As an example let A B, Fig. 26, be a beam 9.6568 units long, or twice the length already found (**47**), and carrying a load of 2 to 1 of length. The length of each cantilever end will now = 2, each point of contrary flexure will be .4142 × 2 = .8284 from the nearest point of support, and the distance between the points of contrary flexure will equal as before twice the length of a cantilever end or 2 × 2 = 4.

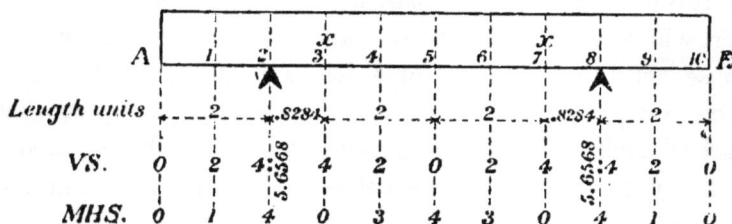

Fig. 26.

Thus the length of the beam is 2 + .8284 + 4 + .8284 + 2 = 9.6568, and the points of contrary flexure are at verticals 3 and 7. Series VS gives the vertical stress at each given vertical. Series MHS gives the moments of horizontal stress at the given verticals in accordance with **27** between verticals 3 and 7, with **42** and **43** from verticals 2 to 3 and 7 to 8, and with **43** between A and vertical 2 and 8 and B.

49. Weight of the Beam, Fig. 26, as a Plate Girder, with End Supports, as compared with Intermediate Supports. Suppose this beam were supported at each end and carried the same load of two to one of length, then the moment of horizontal stress at its central vertical 5 would be 23.3, and the maximum vertical stress would be 9.6, *together* 32.9, whereas, when the supports are placed as shown, Fig. 26, the maximum moment of horizontal stress is 4, and the maximum vertical stress is also 4, *together* 8. Therefore assuming, as in small girders, that the same transverse section is maintained throughout, four equally efficient girders could be supplied under the latter conditions from the amount of material required for one girder of the same length to carry the same load, but supported at each end.

50. Practical Examples. Paddle boards or "floats" of steam vessels, subjected to an uniformly distributed pres-

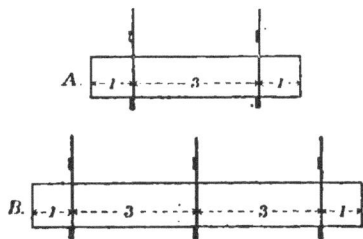

Fig. 27.

sure when in action, are beams of this order when their ends project beyond the paddle wheel arms to which they are attached, the paddle arms being the supports or points of resistance.

Mr. Field's rule* for floats attached to two arms was to divide the length of the float into five equal parts, and to place the arms three apart, thus leaving a projection for each end of the floats of one part beyond the arms. In the case of three arms to each float, the length would be divided into eight equal parts, one arm being at the centre, and the other two each one part from the end of the float (Fig. 27). For as the conditions of stress of the two intermediate spans in the diagram B are precisely the same as those of the one span in diagram A, their relative lengths to that of the cantilever ends will remain the same. This good practical rule leaves the ends or most exposed parts of a float proportionately a small fraction shorter than theory would determine in providing that the moment of stress in the float at an arm of the wheel and at the centre between the arms shall be equal,

* The late eminent member of the firm of Maudslay, Sons, & Field, and formerly President of the Institution of Civil Engineers.

whereas by the given rule the stress at an arm and at the centre between the arms is 1 to $1\frac{1}{4}$.*

For if a be the length of each cantilever end of an uniformly loaded beam, and b half the length of an intermediate span, the stress at a support when compared with that at the centre of a span will be as a^2 is to $b^2 - a^2$. *See* **29, 47.**

The same considerations, both theoretical and practical, are applicable in determining the position of *ledges for doors*, and of *hinges and fastenings* for gates and valves which may have to resist a pressure of wind or water.

Note (*a.*)—Referring to Fig. 25 (**47**), let A B = 2L, S S' = 2l, AS = S'B = l', then L = $l + l'$ and let the load upon each unit of length be regarded as the unit of load.

Let $ff' = 2x$, then the vertical stress at f occasioned by the weight of the load on $ff' = x$, the load upon f S = $l - x$, and the load on the cantilever A S = l', and their respective moments about S will be

$$x\,(l - x), \quad \frac{(l - x)^2}{2}, \quad \text{and} \quad \frac{l'^2}{2}.$$

Therefore the moment of horizontal strain over the support S about S $= \dfrac{l'^2}{2}$ or $= x\,(l - x) + \dfrac{(l - x)^2}{2}$, and that of the horizontal strain at the centre of the beam $= x^2 - \dfrac{x^2}{2} = \dfrac{x^2}{2}$.

But these must be equal

$$\therefore x = l'$$
$$\text{and } \frac{x^2}{2} = lx - x^2 + \frac{l^2}{2} - lx + \frac{x^2}{2}$$
$$\therefore l^2 = 2x^2 = 2l'^2$$
$$\text{and } l = \sqrt{2}\, l'$$
$$L = (1 + \sqrt{2})\, l'$$
$$l = \frac{L}{1 + \sqrt{2}} = (\sqrt{2} - 1)\, L = .4142\, L.$$

That is to say, half the length of the beam being 1, the most efficient position of the supports S and S' is when their relative distance from the extremity A or B is .4142.

And as x has been found $= l'$, or A S in Fig. 25, and 2 being the length of the beam, the distance of ff' from the supports S S' = 1 — .4142 × 2 = .1716 as stated, (**47**).

---

* In a float when regarded as a beam, any theoretic excess of efficiency at a paddle arm would, however, be slightly diminished by the necessary bolt holes.

CHAPTER IV.

51. The static principles which apply to beams extending continuously over two or more spans are further
developments of those which determine the stresses in
beams with cantilever ends (**44, 46, 47, 48**). Of the
whole load, however, upon any one of the spans, the fraction
carried by each support, or the position of the points of
contrary flexure are not so readily found. When this,
however, has been done, a continuous beam of two spans
may be regarded as virtually divided at those points into
three distinct spans,
and a beam of three
spans into five, and
the vertical and
horizontal stresses
may then be easily
ascertained throughout each of these spans constituting the
beam.

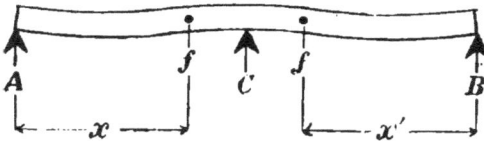

Fig. 28.

Let A B, Fig. 28, be a continuous beam supported at each
end A and B, and also by an intermediate support C, thus
extending over the two spans A C and C B. Now suppose
the two spans each to carry either a concentrated or a distributed load, they will then be deflected downwards by the

load, the horizontal stresses will vanish at the points of contrary flexure *f* and *f* in consequence the reaction of the intermediate support C, and become reversed beyond these points in the same manner as by the central portion of the load in the case of beams with cantilever ends, so that the beam will have a tendency to take a curved form, as shown in the Figure.

In the case of a beam of three continuous spans, Fig. 29, supported at each end A and B and by two intermediate

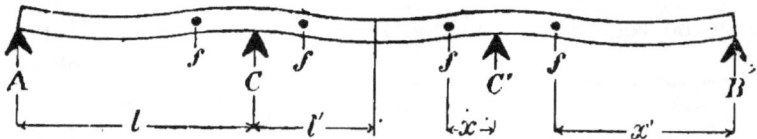

Fig. 29.

supports C and C', there are, or may be, * four points of contrary flexure *f f f f*, two of these being in the central span.

The positions of the points of contrary flexure are dependent upon the relative lengths of the spans, the nature and extent of loading, and the fraction of it transferred to each support by deflection.   These positions and the fraction of the load carried by each support cannot therefore be determined simply by the principle of the lever, but only by the solution of more complicated problems.   The following simple formulæ based upon such solutions are, however, applicable to various conditions of continuous beams.

52. The simplest mode of treating the four following conditions of continuous beams of two spans is to determine by

----

* Not necessarily. If the intermediate supports are very near together there will be no contrary flexure between them.

formulæ (2, 3, 4, and 5) the position of the point of contrary
flexure in each span. When this has been done the beam,
as has just been shown, becomes virtually divided into three
spans. The fraction of the load carried by each of such spans
may then be easily found and the strains at any given verticals
ascertained by simple calculation (**54, 55, 57, 58**).

For Two Equal Spans with Equal or Unequal Central Loads.

Referring to Fig. 28 (**51**).
Let l = A C = C B = the length of each span.
w = the load at the centre of the span A C.
w' = the load at the centre of the span C B.
x = the distance of the point of contrary flexure from A.
x' = the distance of the point of contrary flexure from B.

Then $x = \dfrac{16w}{19w+3w'}l$
$\left.\begin{array}{c}\\[3ex]\end{array}\right\}$
$x' = \dfrac{16w'}{19w'+3w}l$

For equal loads,　　　　　　　(2)
when $w = w'$

$$x = x' = \frac{16}{22}l = \frac{8}{11}l\,*$$

For Two Equal Spans with Equal or Unequal Distributed
Loads.

Let w = the load per linear unit of A C.
w' = the load per linear unit of C B.

* *See* **54, 55.**

Then $x = \dfrac{7\,w\,-\,w'}{8\,w}\,l$ $\left.\right\}$ For equal loads,   (3.)
  when $w = w'$

$x' = \dfrac{7\,w'\,-\,w}{8\,w'}\,l$ $x = x' = \dfrac{6}{8}\,l = \dfrac{3}{4}\,l$ *

For Two Unequal Spans with Equal Central Loads.

Let $l =$ the length of the span A C.

$l' =$ the length of the span C B.

L = A C + C B = the two spans.

Then $x = \dfrac{8\,L\,l^2}{8\,L\,l + 3\,(l^2 + l'^2)}$

$x' = \dfrac{8\,L\,l'^2}{8\,L\,l' + 3\,(l^2 + l'^2)}$ $\left.\right\}$ (4.)

For Two Unequal Spans with an Equally Distributed Load.

Then $x = \dfrac{3\,l^2 + l\,l' - l'^2}{4\,l}$

$x' = \dfrac{3\,l'^2 + l\,l' - l^2}{4\,l'}$ $\left.\right\}$ (5.)

The two following cases of continuous beams of two spans are most simply treated by first finding the portion of the loading taken by each of the end supports A and B. This

---

* *See* Formula (11), and **57, 58.**

having been done the load carried by the central support is known, and consequently the points of contrary flexure. With these data the stresses throughout the beam may be readily determined.

For Two Unequal Spans with Unequal Central Loads.

Let w = the load at the centre of the span A C.

w' = the load at the centre of the span C B.

Then the portion of the load on A $= \dfrac{w}{2} - \dfrac{3\,(w\,l^2 + w'\,l'^2)}{16\,L\,l}$

on B $= \dfrac{w'}{2} - \dfrac{3\,(w'\,l'^2 + w\,l^2)}{16\,L\,l'}$ $\quad$ (6.)

on C $= \dfrac{w + w'}{2} + \dfrac{3\,(w\,l^2 + w'\,l'^2)}{16\,l\,l'}$

For Two Unequal Spans with Unequal Distributed Loads.

Let W = the total load on the span A C.

W' = the total load on the span C B.

Then the portion of the load on A $= \dfrac{W}{2} - \dfrac{W\,l^2 + W'\,l'^2}{8\,L\,l}$

on B $= \dfrac{W'}{2} - \dfrac{W\,l^2 + W'\,l'^2}{8\,L\,l'}$ $\quad$ (7.)

---

53. For Three Spans, the Two Outer Spans being equal, with an equally Distributed Load throughout.

Referring to Fig. 29, (**51.**)

Let $l =$ the length of each of the outer spans A C and C' B.

$2\ l' =$ the length of the central span C C'.

$w =$ the load per unit of length.

Then the load on A or B $= \dfrac{w}{8}\left\{\dfrac{3\ l^3 + 12\ l^2\ l' - 8\ l'^3}{2^2 + 3\ l\,l'}\right\}$

$$\text{on C or C'} = \dfrac{w}{24}\left\{\dfrac{15\ l^3 + 60\ l^2 l' + l l'^2 + 24\ l^3}{2^2 + 3\ l\ l'}\right\} \tag{8.}$$

For Three Equal Spans, or $2l' = l.$

Then the load on A or B $= \dfrac{4}{10}\ w\,l$

$$\text{on C or C'} = \dfrac{11}{10}\ w\,l \tag{9.}$$

With beams of more than three equal spans $x = .28\,l$ may be used for the intermediate spans, and $x' = .8\,l$ for the end spans, as the differences from the true values are so small as to be practically unimportant.

When the Outer Spans are equal, and the Central Span equal to the two, or $l' = l.$

Then the load on A or B $= \dfrac{7}{32}\ w\,l.$

$$\text{on C or C'} = \dfrac{57}{32}\ w\,l. \tag{10.}$$

If $l' = 0$ (reducing the spans to two). *

Then the load on A or $B = \dfrac{3}{8} wl.$

on C and C′ together $= \dfrac{5}{4} wl.$ (11.)

A Continuous Beam of Two Spans with a Load at the Centre of each Span.

54. Vertical Stress. Let A C, Figure 30, be one span of a continuous beam of two equal spans of uniform section, each 22 units long, and one deep; and each loaded at the centre with 16 units, C being the central support dividing the two spans.

Under these conditions the vertical of contrary flexure f is

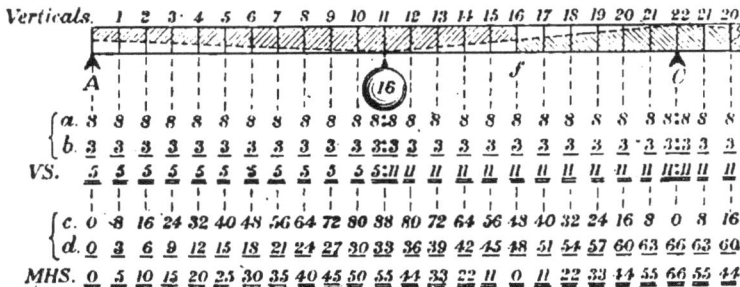

Verticals. 1 2 3 4 5 6 7 8 9 10 11 12 13 14 15 16 17 18 19 20 21 22 21 20

A (16) f C

a. 8 8 8 8 8 8 8 8 8 8 8 8:8 8 8 8 8 8 8 8 8 8:8 8 8

b. 3 3 3 3 3 3 3 3 3 3 3 3:3 3 3 3 3 3 3 3 3 3 3:3 3 3

VS. 5 5 5 5 5 5 5 5 5 5 5 5:11 11 11 11 11 11 11 11 11 11 11:11 11 11

c. 0 8 16 24 32 40 48 56 64 72 80 88 80 72 64 56 48 40 32 24 16 8 0 8 16

d. 0 3 6 9 12 15 18 21 24 27 30 33 36 39 42 45 48 51 54 57 60 63 66 63 60

MHS. 0 5 10 15 20 25 30 35 40 45 50 55 44 33 22 11 0 11 22 33 44 55 66 55 44

Fig. 30.

* *See* also Formula (3.)

at 16 units from the outer support A, and consequently 6 from the central support C, Formula (2), (**52**). A f may therefore be treated, as indicated by hatching in the figure, as a separate beam carrying a load of 16 units at vertical 11, and supported at its two ends A and f. There would thus be a vertical stress of 5 at A and 11 at f (**30**), the 11 at f being continued to C, meeting at that vertical the same amount from the other half of the beam. Now *if the beam were non-continuous* at C, the load of 16 would cause a vertical stress of 8 at A, and 8 also at C, and at all intermediate points between A and C (Series a). But being continuous with a vertical stress of 5 at A, and of 11 at f and at C, it follows that owing to the reaction of the central support (Series b) the vertical stress becomes $8 - 3$ or 5 from A to vertical 11, and $8 + 3$ or 11 from vertical 11 to vertical 22 at the central support C (Series VS).

55. Horizontal Stress. Assuming the span A C to be non-continuous at the central support C, then the moments of horizontal stress due to a load of 16 at vertical 11 are, as given, Fig. 30, Series c, (**24**). But it has been shown that when the two spans are continuous, the reaction of the central support C relieves the support A of $\frac{3}{8}$ of the load carried by it in the former case, and thus Series d gives the moments of horizontal stress due to that reaction. Subtract at each vertical between **A** and f the amount stated in Series d from that in Series c, and as at f the stresses are reversed, subtract Series c from Series d between f and the central support C. We thereby obtain Series MHS, giving the moment of horizontal stress at each relative section of the continuous beam. It will be

F

observed that at the vertical of contrary flexure f the value
of MHS is 0, the factor values (Series c and d) which are
equal, having been deducted one from the other; also that
f C has become a cantilever supported at C and carrying a
load of 11 at its end f, and that the greatest horizontal stress
is 66 over the central support; whereas when the span A C
was assumed to be non-continuous at the central support C,
the greatest horizontal stress was 88 at its central vertical 11.

The stresses will of course be the same respectively at the
same relative verticals of the other equal and equally loaded
span or second half of this continuous beam which is not
shown in the Figure.

56. Comparative Weight of a Beam as a continuous and as two non-continuous plate girders extending over two equal spans and carrying an equal load at the centre of each span.

Firstly, the strength and consequently the weight of
the web at any section will theoretically be directly as the
vertical stress which it has to resist, and referring for example
to Fig. 30 (**54**), Series a and Series VS, it will be seen
that when the girder is non-continuous there is a vertical
stress of 8 units throughout the whole span of 22 units,
whereas for the continuous girder there is a stress of
5 throughout 11 length units and of 11 throughout the
remaining eleven units.

Then $22 \times 8 = 11 \times 5 + 11 \times 11 = 176 \times 2 = 352$
So that the total quantities in each case are equal.

Also the strength and weight of the tables at any section
will likewise in either case vary directly with the horizontal
stress as given per unit of length in Series c and Series
MHS.

Then adding together the numbers in each of those Series we obtain—

for the non continuous girder, Series *c*, $968 \times 2 = 1936$
for the continuous „ MHS, $638 \times 2 = 1276$

Complete the comparison in each case by adding the sum representing the weight of the tables to that representing the web.

Then for the non-continuous girder $1936 + 352 = 2288$
for the continuous „ $1276 + 352 = 1628$
or as 4 to 3 nearly.

Therefore the weight *theoretically* required for the beam when continuous is $\frac{1}{4}$ less than that of two non-continuous beams of the same strength, covering the same spans.

Practically, the weight of girders will of necessity always exceed that determined by theory, and this for a variety of reasons, as will be seen by subsequent examples.

A Continuous Beam of Two Spans with a Distributed Load.

57. Vertical Stress. Let A B, Fig. 31, be a continuous beam divided by a central support C into two equal spans, each 8 units long and 1 deep, each span carrying an equally distributed load of 16 units.

Under these conditions the verticals of contrary flexure *f f'* are at a distance of 6 units from either end of the beam or $\frac{3}{4}$ the length of each span measured from the outer support, Formula (3), (**52**) and the two halves of the continuous beam may be regarded as three distinct beams A *f*, *f f'*, and *f'* B, as indicated by hatching in the figure. The first and third span each carries a distributed load = 12 causing a vertical stress = 6 at A, f, f', and B. The second span $f f'$, conse-

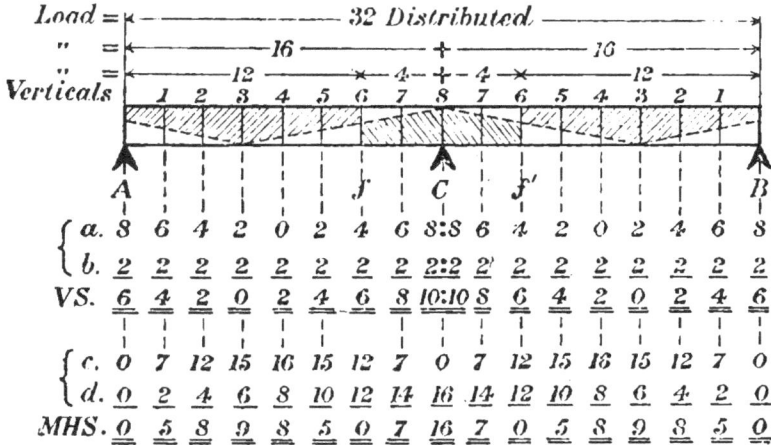

Load =								*32 Distributed*								
" =			16				+				16				
" =		12			4	+	4		12					
Verticals	1	2	3	4	5	6	7	8	7	6	5	4	3	2	1
	A					f		C		f'					B
a.	8	6	4	2	0	2	4	6	8:8	6	4	2	0	2	4	6	8
b.	2	2	2	2	2	2	2	2	2:2	2	2	2	2	2	2	2	2
VS.	6	4	2	0	2	4	6	8	10:10	8	6	4	2	0	2	4	6
c.	0	7	12	15	16	15	12	7	0	7	12	15	16	15	12	7	0
d.	0	2	4	6	8	10	12	14	16	14	12	10	8	6	4	2	0
MHS.	0	5	8	9	8	5	0	7	16	7	0	5	8	9	8	5	0

Fig. 31.

quently carries a load = 6 at each end, and also on each arm 4 units of the distributed load, the load therefore on the central pier C = (6 + 4) × 2 = 20.

Supposing the beam instead of being continuous to be separated at the central vertical 8 into two distinct beams, then Series *a*, Fig. 31, gives (**26**) the vertical stresses for each of the two beams, the stress at each of their ends being 8 units. But it has been shown that in the continuous beam the stress at A or B is 6, and that at the central pier C is 20. As in the previous example, this alteration in the stresses is effected by the reaction, Series *b*, caused by the additional load thrown upon the central pier, by which in this instance

Series *a* becomes reduced by 2 from A and B to the verticals *f* and *f′*, and increased by the same amount between those verticals, Series VS.

58. Horizontal Stress. As in the previous example, the moments of horizontal stress (**27**) in the two spans A C and C B when separated at the central vertical are given in Series *c*, and as it has been shown that the additional amount of vertical reaction due to the central support is 2 at each end of the continuous beam, Series *d* gives the consequent moments of horizontal reaction (**24**) throughout its length. The differences between these two series constitute the Series MHS, or the moments of horizontal stress at each assumed vertical throughout the continuous beam. At the verticals of contrary flexure they are nil, (**55**).

59. Comparative Weight of a Beam as a continuous and as two non-continuous plate girders of two equal spans carrying an equally distributed load. We assume as in Art. **56** the weight of each member of a beam of a given span to be directly as the stresses its members have to resist.

Thus in this instance for the web multiply the mean of the stresses given in Series *a*, Fig. 31, and the mean of those in Series VS, by the length in each case over which each mean thus taken extends.

Then for the web of the non-continuous beam (Series *a*)

$$4 \times 8 = 32 \times 2 = 64.$$

for the web of the continuous beam (Series VS)

$$(3 \times 6) + (8 \times 2) = 34 \times 2 = 68.$$

For the tables add together the numbers in Series *c*, and also those given at each unit of length in Series MHS.

Then for the non-continuous beam (Series *c*) $84 \times 2 = 168$.

for the continuous beam (Series MHS) $50 \times 2 = 100$.

Adding together these results respectively we obtain—

for the non-continuous beam $168 + 64 = 232$,

for the continuous beam $\qquad 100 + 68 = 168$,

or as 4 to 3 nearly.

Or as in the example given in **56** the weight required for the continuous beam is about $\frac{1}{4}$ less than that of two similar non-continuous beams of the same strength.

60. Comparative Deflection of a Beam when non-continuous and continuous.

Assume the beam, Fig. 31, to consist of a top and bottom table, connected by a web, and whether non-continuous or continuous to be of uniform depth and of uniform breadth throughout, but to have its parts so proportioned in thickness as to meet all stresses with an equal resistance per unit of section. Then the linear extension or compression caused by the load might, under each condition, be represented by 8, the number of units in the length of each actual span.

But continuity has been found to divide the beam into three virtual spans, the two outer of which are each 6, and the central span 4 units long; note also that the *versed sines* of comparatively flat segmental curves vary as the squares of their chords.

Therefore, under these conditions, the comparative deflection of each actual span may be represented in the following way :—

When non-continuous $8^2 \qquad = 64$

When continuous $\qquad 6^2 + 4^2 = 52$

or as 5 to 4.

Therefore the maximum deflection will be $\frac{1}{5}$th less when the two spans are covered by a continuous beam than it would be were the same load carried by two equally efficient but separate beams.

It has been shown, therefore, that the ratios of weight and deflection are in favour of the continuous beam. In practice, however, the ratio of weight would be somewhat less favourable, for, in order to insure lateral rigidity, the tables could not be reduced to nothing at the verticals of contrary flexure, as would be the case if proportioned to the horizontal stresses.

CHAPTER V.

61. Experiments on the strength of materials and their behaviour under stress of all kinds gradually increased to the moment of complete fracture, have from time to time been carefully carried out by experienced and accurate observers, assisted by the most advanced and complete appliances.

The results of many series of such experiments have been carefully tabulated, averaged, and published. Moreover the process of simply testing the ultimate transverse and tensile strength of iron and steel is an every-day process, necessary for the purpose of ascertaining whether the quality of material is equal to stipulated standards of efficiency. Very many valuable data are therefore available in determining proper standards with practically sufficient accuracy.

The transverse strength of any constructive material is usually experimentally determined by placing a succession of bars of given scantlings* upon or against two points of support or resistance, adjusted to a given distance apart. Pressure is then applied separately to each bar at a point midway between the points of resistance, and in a direction at right angles to its length.

The pressure may be effected simply by the gradual appli-

---

* A term used for dimensions other than length.

cation of suspended known weights until the bar is broken. But it will be obvious that the gradual and at the same time self-registering pressure obtainable by the use of a hydraulic ram and pressure gauge is preferable for ordinary tests, especially when the condition of the bar is approaching fracture. Further, by the use of such an apparatus, tests may be made with much greater expedition.

On the other hand, although the gauge if correct will give the pressure of the water within the cylinder of the ram, the actual stress to which the object is being subjected is the pressure indicated by the gauge multiplied by the area of the ram, less the friction of its hydraulic packing.

The most accurate mode of testing is therefore with a lever machine,* in which friction is reduced to a minimum by the fulcrums of the levers being all fitted on the knife-edge scale beam or steel-yard principle. The weight causing the transverse stress on the object to be tested is moved by a hand-wheel and band along a graduated lever, so that the exact amount of the stress upon the test bar can be at once read and noted at any period of the operation.

Tensile strength is ascertained by securing each end of the bar to be tested in the testing machine, by means of which it may be placed in a state of tension, and the stress gradually increased and continuously registered up to the required test, or until the bar is torn asunder.

Means are also provided for accurately measuring the extension and the deflection of a bar during a test.

---

* As at Messrs. David Kirkaldy and Son's establishment and Museum of Experimental and Accidental Fractures, 99, Southwark Street, London, by whose machine materials may also be tested under tensile, compressive, shearing, and twisting stress.

62. Coefficients of Efficiency. The figures given
in columns c and c', n and n' of the following table are computed at not more than one-fourth to one-fifth of the average
ultimate strength of steel, iron, oak, and fir, of the descriptions generally used in the construction of beams, girders,
and other structures. These are termed *coefficients of efficiency*.

TABLE 1.

Nature of Strain.		Material.	Bar 1 in.by 1 in. Square.	Bar 1 in. Diameter.
			c Tons.	c' Tons.
Tensile compressive and shearing	}	Rolled Steel	6.50	5.20
		„ Iron	4.50	3.53
Tensile and shearing · · ·		Cast Iron	1.40	1.10
Compressive · · · · ·		„ „	6.00	4.70
Tensile and compressive · ·		Oak	.50	.39
„ „ · ·		Fir	.40	.31
			n	n' .
	(	Rolled Steel	.60	.374
		„ Iron	.40	.245
		Cast Iron	*.175	.112
Transverse; Bars one foot between the supports and loaded at their centre · · · · ·	{		Cwt.	Cwt.
		Oak	1.00	.60
		Fir	.80	.50
	(	Landings Hard York }	.10	

The *ultimate* compressive strength of wrought-iron plates
and bars is when compared with their tensile strength about as

* = 3¼ cwt., or for small work, say, 490 lbs., for machinery 200 lbs.

4 to 5. But the same co-efficient of *efficiency* may be used in either case, and especially for girder work. For rivet and bolt holes diminish the efficiency of sections in tension, but not to the same extent, that of those in compression, inasmuch as the rivets or bolts fill up the holes and so resist compression. It is usual therefore to simplify and expedite such work by providing the same *net* sectional area for the resistance of the same amounts of stress, whether tensile or compressive.

As the tensile strength of iron plates is 10 per cent. more lengthways in the direction of the fibre, or that in which they were rolled, than the strength crossways, they should be placed lengthways when used for resisting that stress.

Shearing efficiency is equal to tensile in rolled iron plates and bars.

In steel compressive efficiency is fully equal to tensile, and although its *ultimate* shearing strength is one quarter less than its tensile, the coefficient $6\frac{1}{2}$ tons given in the table is practically a well-covered allowance of safe shearing efficiency.

Engines and machinery with members in motion are subject to somewhat indefinite abnormal strains. The maximum calculated stresses therefore which their parts may have to resist per square inch of sectional area should in no case exceed one-half of those allowed for ordinary fixed structures, while the materials used in their construction should be of the best descriptions.

63. Crushing Resistance of Materials. The results of the various published series of experiments on the resistance offered by building materials to crushing stress vary in many instances very considerably, and moreover the strength of a pier or wall of brick or stone decreases with its height, by reason of increased tendency to bend. No absolutely definite coefficients of safety can therefore be laid down.

But any amount of stress not exceeding that given in each case in the following Table will, under ordinary constructive conditions and requirements, be practically well within the limits of safety.

TABLE 2.

Safe Loads for Earth and Materials.

	Per Square Foot.
Upon Firm Earth	1½ Ton.
„ Hard Gravel or Clay	3 Tons.
For Lias Lime Concrete, 6 of Ballast, and 1 of Cement .	6 „
„ Portland Cement Concrete, 8 of Ballast, and 1 of Cement	7 „
„ Brickwork, ordinary	6 „
„ „ in Lias Lime Mortar	7 „
„ „ in Portland Cement	8 „
„ Granite, Cornish or Aberdeen	30 „
„ Portland Stone	20 „
„ Hard York Stone and Landings	20 „

For important works it is usual to determine the practically effective supporting strength of the materials intended to be used, by special experiments upon samples tested under conditions relative to the proposed work.

WEIGHT OF MATERIALS.

64. In taking out quantities from drawings either for the purpose of calculating stresses caused by the weight of a structure or for making estimates of cost, a ready and expeditious mode of proceeding with iron or steel is to bring castings, plates, angle, and T iron into square superficial feet measure, termed feet super, arranged according to thicknesses.* The weights of the usual thicknesses are given in the following Tables :—

TABLE 3.

Cast Iron.

1 Foot cube = 4 cwt., therefore 5 feet cube = 1 ton, and

$$\frac{Thickness\ in\ inches}{3} = \text{cwt. per foot super, thus :—}$$

One foot super
$$
\begin{cases}
3 \text{ inches thick} = 1 \text{ cwt.} \\
2 \quad ,, \quad ,, = \tfrac{2}{3} \;,, \\
1\tfrac{1}{2} \quad ,, \quad ,, = \tfrac{1}{2} \;,, \\
1\tfrac{1}{4} \quad ,, \quad ,, = \tfrac{5}{12} \;,, \\
1 \quad ,, \quad ,, = \tfrac{1}{3} \;,, \\
\tfrac{7}{8} \quad ,, \quad ,, = \tfrac{7}{24} \;,, \\
\tfrac{3}{4} \quad ,, \quad ,, = \tfrac{1}{4} \;,, \\
\tfrac{5}{8} \quad ,, \quad ,, = \tfrac{5}{24} \;,, \\
\tfrac{1}{2} \quad ,, \quad ,, = \tfrac{1}{6} \;,, \\
\tfrac{3}{8} \quad ,, \quad ,, = \tfrac{1}{8} \;,, \\
\tfrac{1}{4} \quad ,, \quad ,, = \tfrac{1}{12} \;,, \\
\tfrac{1}{8} \quad ,, \quad ,, = \tfrac{1}{24} \;,,
\end{cases}
$$

* Although in many handbooks tables may be found of the weight of angle and T iron per foot run.

As dimensions are usually given in feet, inches, halves, quarters, and eighths, *duodecimals* are far preferable to decimals in taking out quantities, because with the former far fewer figures are required, and $\frac{1}{3}$ or $\frac{2}{3}$ cannot be expressed by simple decimals.

Plate Iron.

One foot cube = 480 lbs., therefore $4\frac{2}{3}$ feet cube = 1 ton, and—

1 inch thick = 40 lbs. per square foot.

$\frac{3}{4}$,, ,, = 30 ,, ,,

$\frac{1}{2}$,, ,, = 20 ,, ,,

$\frac{1}{4}$,, ,, = 10 ,, ,,

$\frac{1}{8}$,, ,, = 5 ,, ,,

In girder work 2 or 3 per cent. added to the weight for rivet heads would generally be sufficient, but it is usual to add 5 per cent.

Steel.

Add 1 per cent. to the weight of plate iron.

Timber.

Per foot cube   ...   Oak, .5 cwt.

,,   ,,   ...   Fir, .4 ,,

65. Data for estimating the Loading of Structures.

TABLE 4.

Weight of Materials, &c.

Per foot cube.	Brickwork in lime or cement	1.00 cwt.
,,	,,	Gravel or ballast . .	1.25 ,,
,,	,,	Portland cement concrete .	1.25 ,,
,,	,,	York or Portland stone .	1.40 ,,
,,	,,	Granite	1.50 ,,

Per square foot 6 ins. thick York or Portland Paving .70 cwt.

,,　　　,,　　3 ins. . . . . . .35 ,,

Per square 10 × 10 = 100 square feet plain Tiling　12.00 ,,

,,　　,,　　,,　　,,　　,,　　pan Tiling　8.00 ,,

,,　　,,　　,,　　,,　　,,　　Slating .　7.00 ,,

,,　　,,　　¾-inch deal boarding　. .　2.50 ,,

Per square foot. A Crowd of persons *　.　1.50 ,,

,,　　,,　　House floors, persons, and
　　　　　　　　furniture . . .　1.50 ,,

,,　　,,　　Store and Warehouse floors,
　　　　　　　　loading . . .　2 to 4 ,,

,,　　,,　　Wind—probable maximum †⎱ 40 lbs.
　　　　　　　　pressure in the British Isles ⎰

Per foot cube fresh Water . . . .　62.5 lbs.

36 feet　,,　　,,　,,　. . . .　1 ton + 10 lbs.

Per foot　,,　sea Water . . . .　64 lbs.

35 feet　,,　　,,　,,　. . . .　1 ton.

Per square foot. A fall of Snow per each inch ⎱ .5 lbs.
of depth before becoming consolidated . ⎰

---

* Careful experiments show that a crowd of men of average weight closely packed weigh 1¼ cwt. per square foot of the area covered. The adoption of 1¼ cwt. per square foot in planning a house will therefore leave a margin of strength for further contingencies.

† By a *Board of Trade regulation* for the stability of viaducts and high bridges, " the work must be such as will provide for a wind pressure of 56 lbs. on the square foot." This pressure was fully provided for in the Forth Bridge by Sir Benjamin Baker. Special trussing was provided at the Crystal Palace, Sydenham, for a wind pressure of 25 lbs. per square foot, and even with that there would be 66 tons pressure upon the area of the semi-circular part, of an end of the large transept, which has now stood for over five-and-thirty years. The *ultimate* strength of this trussing is, however, four times that provided, so that a hurricane of 100 lbs. per square foot would have to occur in order to cause any serious failure.

Bridges carrying public roads should be designed to carry the following loads:

The weight of the structure platform and roadway together with a distributed load of $1\frac{1}{2}$ cwt. per foot super over the entire platform. Also a maximum *concentrated rolling load* upon any part of the platform, but not in addition to the distributed load.

For the latter, in the case of town roads, a load of 32 tons on four wheels including the trolley, or 8 tons on each wheel, would be a practically good allowance, assuming the wheels to be 10 feet apart from centre to centre longitudinally, and 4 feet apart from centre to centre transversely.* †

Half the above weight or 4 tons on each wheel taken at the same distances apart would be a sufficient allowance for traction engines in the case of country roads.

For *bridges carrying railways*, although no absolute rule has been laid down by the Board of Trade, the test rolling loads for each line or pair of rails usually adopted have been per foot run about $1\frac{1}{2}$ ton up to a 50 feet span, $1\frac{1}{4}$ ton from 50 to 100 feet span, and 1 ton from 100 to 150 feet span.

---

* The load adopted by Sir Joseph Bazalgette 15 years since.

† Messrs. Maudslay, Sons, and Field frequently carry from 32 to 35 tons on their large four-wheel trolley. They have had 43 tons upon it, to which add $5\frac{1}{2}$ tons for the trolley and 1 ton for chain, making in all $49\frac{1}{2}$ tons or $12\frac{3}{8}$ tons on each wheel. These are, however, for roads or streets, concentrated loads of unusual weight.

CHAPTER VI.

66. The Transverse Strength of Solid Rectangular Beams or bars varies directly as their breadths and as the square of their depths (**19, 20**), and that of cylindrical as the cubes of their diameters, and for any section inversely as their lengths. The latter factor is wholly one of leverage. Thus assuming a beam to be supported at each end, and its transverse section to be uniform throughout its length, the stress at any segmental transverse section would vary in direct proportion with the length of the beam.

It therefore follows that as the transverse strength of plain solid sections of constructive materials has been determined by direct tests (**61**), the sections of such rectangular and round beams required for given loads, and the loads which will be safely carried by beams of such sections, may be readily found by the use of the following formulæ.

Let b = the breadth $\Big\}$ in inches.
d = depth

l = length in feet between supports.
n = coefficient for rectangular bars $\Big\}$ Table 1 (**62**).
n' = „ circular „
w = load at centre of span.

G

For Central Loads.

67. Required the load that a beam of oak 12 feet long between the supports and 12 inches by 12 inches section will carry at its centre.

$$\text{Then } w = \frac{b\,d^2\,n}{l} \tag{12.}$$

$$= \frac{12 \times 12^2 \times 1}{12} = 144 \text{ cwt.}$$

Required the breadth of a beam of fir 15 feet long between the supports and 10 inches deep to carry a load of 40 cwt. at its centre.

$$\text{Then } b = \frac{w\,l}{d^2\,n} \tag{13.}$$

$$= \frac{40 \times 15}{10^2 \times .80} = 7\tfrac{1}{2} \text{ inches}$$

Required the depth of a bar of cast iron 4 feet long between the supports and $2\tfrac{1}{2}$ inches broad to carry a load of $1\tfrac{3}{4}$ tons at its centre.

$$\text{Then } d = \sqrt{\frac{w\,l}{b\,n}} \tag{14.}$$

$$= \sqrt{\frac{1\tfrac{3}{4} \times 4}{2.5 \times .175}} = 4 \text{ inches.}$$

For Distributed Loads.

68. It has been shown (**27**) that if a beam will carry w at its centre, it will carry $2w$ evenly distributed over its span. Formulæ (12), (13). (14) become therefore modified in the following manner for distributed loads :—

Let $w' =$ the weight of a distributed load.

To find the total safe distributed load for a beam.

$$\text{Then } w' = \frac{2\,b\,d^2\,n}{l} \tag{15.}$$

The length, depth, and loading being given to find the efficient breadth of a beam to carry a total distributed load.

$$\text{Then } b = \frac{w'\,l}{2\,d^2\,n} \tag{16.}$$

The length, breadth, and loading being given to find the efficient depth of a beam to carry a total distributed load.

$$\text{Then } d = \sqrt{\frac{w'\,l}{2\,b\,n}} \tag{17.}$$

For Round Beams.

Let d = the diameter.

69. To find the load that a bar of rolled iron 4 feet long between the supports and 4 inches diameter will carry at its centre.

$$\text{Then } w = \frac{d^3\,n'}{l} \tag{18.}$$

$$= \frac{4^3 \times .245}{4} = 3.92 \text{ tons}$$

Required the diameter of a beam of cast iron 8 feet long between the supports to carry a load of 14 tons at its centre.

$$\text{Then } d = \sqrt[3]{\frac{w\,l}{n'}} \tag{19.}$$

$$= \sqrt[3]{\frac{14 \times 8}{.112}} = 10 \text{ inches}$$

For round beams with *distributed* loads w', substitute for Equation (18) $w' = \dfrac{2\,d^3 n'}{l}$, and for Equation (19) $d = \sqrt[3]{\dfrac{w\,l}{2\,n'}}$.

70. The use of the formulæ given (**67, 68, 69**) may be readily extended by noting the following conditions :—

When supported at its centre a beam will carry at each end one-half of the safe central load (**25**), and per unit of length the same distributed load as when supported at each end (**29**).

As a cantilever a beam of any given effective length will carry at its salient end one-fourth of the central load, and also per unit of length one-fourth of the distributed load it will carry when supported at each end (**43**).

With a load at any given vertical in the span transverse stress will be greatest at that vertical, and will vary as the multiple of the separate lengths of the segments into which it may divide the length of the span (**31**).

Or multiply the length of one segment by that of the other and divide the product by one-fourth the length of the whole span, and the quotient will be the effective length with the load placed at its centre.

Thus if l' l'' be the length of the segments, the load carried will be inversely as l the length of the beam.

$$\text{When } l = \frac{4 \; l' \; l''}{l}$$

---

71. The Nature of Transverse Strength was the subject of careful research by Mr. Robert Stephenson, by whose extensive experiments and the deductions to which they led the knowledge of it was much advanced. *

The investigations of Mr. Stephenson were followed by those of Mr. W. H. Barlow, who published an account of

---

* " The Britannia and Conway Bridges," by Edwin Clark, 1850.

a series of elaborately accurate experiments and of the deductions which he based upon them.*

The results of these labours have since been practically elucidated and extended by Sir Benjamin Baker. †

Before Mr. Barlow's solution of this problem two elements only of transverse strength were recognised as existing in a beam, the one being the resistance it is capable of offering to direct extension and the other its resistance to compression. But the efficiency of these resistances was found quite inadequate to account for the amount of strength practically evinced by *solid beams* when measured by or compared with the results of experiments on the direct tensile and compressive strength of materials. In this comparison the neutral axis was assumed as passing through the centre of gravity of a beam. But in order to obtain a moment of horizontal tensile resistance sufficient to account for the strength of a solid cast-iron beam the neutral axis would have to be at or above the top of its transverse section.

That the neutral axis passes through the centre of gravity of a beam of any form of section whatever, provided all the forces acting on the beam are applied in a direction perpendicular to that axis, and that the limits of perfect elasticity are not exceeded, had already been mathematically determined. The consideration of these facts led Mr. Barlow to conclude that another cause existed for the amount of transverse strength as exhibited by solid beams, that the particles of a bar of iron when under transverse stress, besides being in tension below the neutral axis and in compression above it, must be subject to a differential molecular lateral movement among themselves,

* " The Philosophical Transactions of the Royal Society," 1855 and 1857.
† " On the Strength of Beams, Columns, and Arches," by B. Baker, 1870.

and that this movement would necessarily bring into action a corresponding amount of resisting energy, which he named " the resistance of flexure."

Thus referring to **18** and Fig. 6, it will be evident that in a beam subjected to transverse stress, horizontal stress commences at the neutral plane and then increases in a direct ratio to a maximum at the top and bottom of the beam. As the stress increases so will the differential molecular movement of the particles of the beam and their resistance to that movement also increase.

Now there can be no such movement among the particles of a bar under direct tensile or compressive stress, equally affecting the whole of the section strained, for its strength then entirely depends upon the direct resistance it is capable of offering to either of those stresses.

But in the case of bars under transverse stress, their strength then depends not alone upon the resistance they are capable of offering to direct tension and compression, but further to that which is due to their resistance to *bending or flexure.* For Mr. Barlow's experiments showed that the calculated tensile stress per square inch at the lower side of cast-iron bars subjected to transverse stress exceeded, before rupture took place, that which experiments on their extension by direct tension proved them to be capable of resisting in the ratio of nearly two to one. This difference was too great to admit of the old workshop notion being any longer entertained that it is caused by the superior strength of the skin. Besides, in a direct tensile test, the resistance of the whole of the skin is brought into operation, whereas in a square bar under transverse stress one-half only of the skin is in tension, the remainder being in compression. But in a cast-iron bar under transverse stress there is a considerable excess of strength in the part in compression over that in tension quite independent of the resistance of flexure. Besides, experiments on the transverse strength of bars with the skin removed have shown that the skin itself practically possesses no superior strength.

It may be open to question whether this increased resistance is or is not attributable to the *lateral* action of the particles, but in any case it will require to be taken into consideration in the solution of questions relating to transverse strength. *

STRENGTH ELEMENTS IN SOLID BEAMS.

The three following examples illustrate the *theory* of the elements of transverse strength briefly delineated in the last Article.

Practically, however, the efficiency of solid beams is best dealt with by formulæ based upon actual strength tests (**66, 67, 68, 69**) in use long before any real advance had been made in the *theory* of transverse strength.

72. Given a Cast-iron Bar squarely strained

1 foot 6 inches long between the supports, and $1\frac{1}{2}$ in. $\times$ $1\frac{1}{2}$ in. section, to find the breaking load or weight, $b\,w$, when carried by the bar at the centre of the span.

Referring to Table 1 (**62**), take $b\,w$ the ultimate transverse strength of cast iron as $.175 \times 5 = .875$ ton.

Then by Equation (12) (**67**),

$$b\,w = \frac{b\,d^2\,5\,n}{l} = \frac{1\frac{1}{2} \times (1\frac{1}{2})^2 \times .875}{1.5} = 1.969 \text{ tons.}$$

* On the subject of flexure, see an article on "Elasticity" by Sir William Thomson (Lord Kelvin) (*Ency. Brit.*, vol. vii.). See also an article by Professor Ewing on the "Strength of Materials" (*Ibid.*, vol. xxii.).

To resolve the strength elements, let A B C D, Fig. 32,

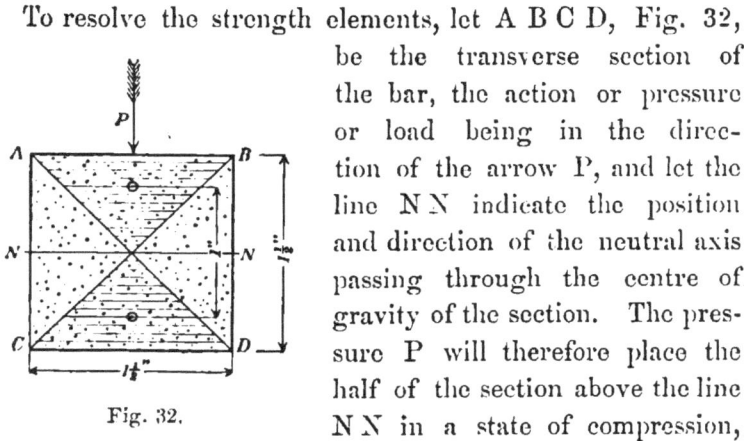

Fig. 32.

be the transverse section of the bar, the action or pressure or load being in the direction of the arrow P, and let the line N N indicate the position and direction of the neutral axis passing through the centre of gravity of the section. The pressure P will therefore place the half of the section above the line N N in a state of compression, and the remaining half below that line in a state of tension (**18**), Fig. 6.

Now the resistance that cast iron is capable of presenting to compression when compared with that which it is capable of offering to tension is $4\frac{1}{4}$ to 1. Therefore as the lower half of the section is in tension, in that part will fracture first commence. Draw the diagonals A D and B C, dividing the diagrams into triangles. Then as the horizontal resistance to stress commencing at the neutral axis N N will increase directly as the distance of any point from that line, it is evident that if the line C D at the maximum distance from N N be taken to represent the tensile resistance of the material per square inch at that line, the length of each of the other horizontal lines drawn across the bottom triangle at intermediate distances will also relatively represent that resistance at each such line. Therefore, the area of that triangle multiplied by the resistance represented by the length of its base C D will be the measure (m) of the ultimate tensile resistance of the section A B C D.

In the same way, the area of the top triangle will represent the measure of the resistance of the section to compression. In the example, the area of each triangle $= \dfrac{1\frac{1}{2}^2}{4}$.

It therefore follows that, referring again to Table 1, and taking $1.4 \times 5 = 7$ tons as the ultimate tensile strength of cast iron, the ultimate tensile resistance (m) of the section

$$= \frac{1\frac{1}{2}^2 \times 7}{4} = 3.9375 \text{ tons.}$$

Now the centre of tensile resistance is the centre of gravity of the bottom triangle, and the centre of resistance to compression is the centre of gravity of the top triangle. These two points are $1\frac{1}{2} \times \frac{2}{3} = 1$ inch apart, Fig. 32, and therefore the effective depth of the section is one inch. Or for rectangular beams, if a be the area of the section, $\dfrac{a}{4}$ is the effective area of resistance, and c being the coefficient of tensile strength $5c$ may be taken as the ultimate tensile resistance.

$$\text{then } m = \frac{5}{4} \, a \, c.$$

and if d be the depth of the beam, $\dfrac{2}{3} d = $ the effective depth.

Having thus determined the ultimate tensile resistance and the effective depth of the section, the multiple of these is the moment of resistance, and the next step is to find the weight of a load placed at the centre of the span which will cause an equal and opposite stress.

Assume half the length of the span and the effective depth of the section to be a bent lever (**24**), and—

Let $m = $ the ultimate tensile resistance (for the given
section $= 3.9375$ tons)
$d = $ the effective depth
$l = $ the length of the span $\Big\}$ in inches
$w = $ the required load

$$\text{Then } w = \frac{* \, 4 \, m \, d}{l} \tag{20}$$

$$= \frac{4 \times 3.9375 \times 1}{18} = .875 \text{ ton}$$

But it has been shown by Equation (12) that given the ultimate strength of cast iron as determined by actual experiments, the breaking load $b \, w$ is 1.969 tons. Now by Equation (20), $w = .875$ tons only, and therefore the resistance of flexure must equal the difference, or

$$1.969 - .875 = 1.094 \text{ tons.}$$

The tensile resistance being .875 and that of flexure 1.094 tons, the former is to the latter as 1 to $1\frac{1}{4}$, or the average of the results of Mr. Barlow's and Mr. Hodgkinson's investigations.

Required the safe central load for a rectangular solid beam squarely strained.

Let a = the area of the section . . ⎫
   a' = effective area of resistance = $\frac{1}{4} \, a$ ⎪
   d = depth of section . . . ⎬ in inches.
   d' = effective depth of section = $\frac{2}{3} \, d$ ⎪
   l = length of span . . . . ⎭
   c = tensile coefficient (**62**, Table 1).

† ⎧ p = resistance of flexure = $1\frac{1}{4} \, c$ for cast iron.
  ⎨ „ =   „   „ = .70 c for rolled steel.
  ⎩ „ =   „   „ = .56 c for rolled iron.

   $F = c + p.$
   w = the safe load at the centre of the span.

$$\text{Then } w = \frac{4 \, a' \, d' \, F}{l}$$

* See Formula (22) (**76**), in which a and c take the place of m.

† In this and subsequent examples the values of p are derived from experiments and investigations made by Barlow, Fairbairn, and Hodgkinson.

73. Given a Cast-iron Square Bar diagonally strained to find the breaking load when applied at the centre of the span.

Let A B D E, Fig. 33, be the section of a bar of the same dimensions as Fig. 32, namely, $1\frac{1}{2}$ in. × $1\frac{1}{2}$ in. and 18 inches in length between the supports, but strained by the load in the direction of the vertical diagonal A E as indicated by the arrow P. The horizontal diagonal B D passing through the centre of gravity of the section at C will then coincide with the position and direction of the neutral axis.

Now in this as in the last example, the lower half of the section will be in tension, and the greatest tensile stress per unit of area will be at the bottom angle E, so therefore at that point would fracture theoretically commence.

Divide the half C D of the diagonal B D into any number, say 8 units of width, and the lower half

Fig. 33.

C E of the diagonal A E into an equal number or 8 units. Place the units of width in Series *a*, and the units of depth in Series *b*, multiply together the relative factors, and place the results in Series *c*.

Let E D′ = C D, and D D′ = C E, then assuming the section of the bar to have extended to D′, E D′ multiplied by half the depth C E, or 8 × 8 = 64, will represent the tensile efficiency on the line E D′, so also will

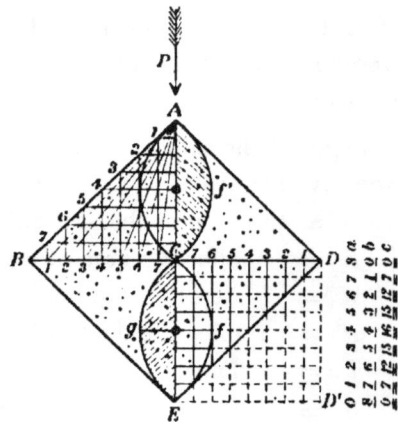

the numbers in Series c represent the proportionate tensile efficiency of the bar at each of the sectional widths to which they respectively refer.

From the line C E let ordinates be drawn parallel to C D, and each proportioned to it in length as the relative number in Series c is to 64. Draw through their extremities the curve C f E, and in like manner on C E the same curve reversed C g E.

These two lower curves inclose the area of equal tensile resistance of the section when submitted to transverse diagonal stress and the two similar curves drawn in the upper half of the section inclose the same area of compressive efficiency.

Now the resulting curve is a parabola, and the area of a parabola is $\frac{2}{3}$ of that of the circumscribed rectangle as may be found by the given ordinates. Therefore taking the units of division as shown in the figure the area of the latter = $(8 \times 2) \times 2 = 32$, and the parabolic area = $32 \times \frac{2}{3} = 21\frac{1}{3}$.

But the whole area of the bar = $8 \times 8 \times 2 = 128$ units, therefore the resolved area of maximum tensile efficiency is to the whole area as 1 to 6, or as may be seen by the Fig. $\frac{2}{3}$ of $\frac{1}{4} = \frac{1}{3}$ of $\frac{1}{2} = \frac{1}{6}$ of the whole section.

Therefore as the bar is $1\frac{1}{2} \times 1\frac{1}{2}$ in. = 2.25 in. area the area of ultimate resistance is $\dfrac{2.25}{6} = .375$ inches.

Taking as in the last example 7 tons per square inch as the ultimate tensile strength of cast iron, $.375 \times 7 = 2.625$, or m the measure of tensile resistance of the section.

It is evident that the effective depth of the bar is the distance $f'f$ measured from centre to centre of the effective areas of resistance, or $\dfrac{\sqrt{1\frac{1}{2}^2 \times 2}}{2} = 1.06$ inches.

Therefore Equation (20), (**72**) to find the weight of a central load w which would cause a moment of stress at the centre of the span equal to the ultimate tensile moment of resistance,

$$w = \frac{4\,m\,d}{l} = \frac{4 \times 2.625 \times 1.06}{18} = .618 \text{ tons.}$$

In the case of square bars strained diagonally the ratio of their resistance to flexure to that of tension has been found to be for cast iron $1\frac{1}{2}$ to 1.

Therefore as $.618 \times 1\frac{1}{2} = .927$, the ultimate load on the centre will be $.618 + .927 = 1.545$ ton. Now by experiment when squarely strained the bar broke with 1.969 ton, therefore if the ultimate strength of a cast-iron bar when squarely strained $= 1$, it will when diagonally strained $= \frac{1.545}{1.969} = .783$, or say $\frac{3}{4}$.

Therefore Formula (12), (**67**) modified is applicable, assuming b and d to be each a side of the square bar when diagonally strained, and l the length of the span in feet.

$$\text{Thus } w = \frac{3\,b\,d^2\,n}{4\,l} \text{ or } \frac{3\,d^3\,n}{4\,l}$$

The area of equal resistance per unit of efficiency may be *geometrically* determined in the following manner.

Divide the lines A C and B C into any equal number of units, say 8 as before, and from the points thus obtained in A C draw horizontal lines to the side A B of the bar.

Then these lines will represent by their respective lengths either as may be required the direct tensile or the compressive strength of the section at each plane taken. From the points obtained in like manner in the line B C, draw vertical lines meeting at the side B A of the bar the horizontal lines already drawn. Then these vertical lines will represent the distances of the several horizontal planes of strength, measured from the neutral axis B D. Draw diagonal lines from each point taken in B C to the angle at A. Point the intersection of each horizontal line with the diagonal numbered the same in the figure, and a line drawn through the points thus obtained will be the parabolic curve defining as before the area of equal resistance.

Required the safe central load w for a solid square beam diagonally strained.

With the following exceptions, the symbols have the same value as those given (**72**).

Let $a' =$ the effective area of resistance $= \dfrac{1}{6}\, a$

$\quad d \ \ = \quad$ width of one side of the beam $\qquad$ in

$\quad d' \ = \quad$ effective depth of section $= \dfrac{\sqrt{2\,d^{2}}}{2}$ $\quad$ inches.

$\quad p \ \ = \quad$ resistance of flexure $= 1\tfrac{1}{2}\, c$ for cast iron.

$\quad ,, \ = \qquad ,, \qquad\quad ,, \quad = .9\, c$ for rolled steel.

$\quad ,, \ = \qquad ,, \qquad\quad ,, \quad = .8\, c$ for rolled iron.

$$\text{Then } w = \frac{4\,a'\,d'\,F}{l}$$

---

74. Given a Rolled Round Iron Bar 2 feet long between the supports, and 2 inches diameter to find the safe load w when carried by the bar at the centre of the span.

Referring to Table 1 (**62**) the safe transverse strength w' of a rolled iron bar 1 inch diameter and 1 foot long = .245 ton, therefore by Formula (18) (**69**),

$$w = \frac{d^3\, n'}{l} = \frac{2^3 \times .245}{2} = .98 \text{ tons.}$$

To proceed as before in resolving the strength elements of this bar, let Fig. 34 represent its transverse section. Then with the load acting in the direction of the arrow P, the half of the whole section of the bar above the neutral axis N N will be in compression, and the other half below that line will be in tension.

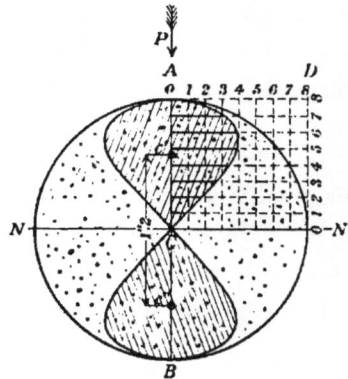

The curves shown in Fig. 34 having been formed as described in the last example,

Fig. 34.

inclose two areas, each of which (a') in this instance = .64 inch, the lower one being the area of equal tensile and the upper one the area of equal compressive resistance.

Although the *ultimate* resistance of rolled iron to compression as compared with that which it offers to tension has been assumed to be about 4 to 5, (**62**) its effective resistance within limits of perfect elasticity to those two stresses may be taken as equal. Besides, this material and also rolled steel has a *tendency* to be " set up," and its sectional area thus increased by compression, whereas it is "drawn out" and diminished by tension. Practically therefore the *efficient* resistance to compression of rolled bars under transverse strain may be taken as equal to their resistance to tension.

Now the effective depth (d') of the section is 1.2 inch, Fig. 34, and the tensile efficiency (c) per square inch of rolled iron = 4.5 tons, Table 1, (**62**).

Therefore, in supporting a central load w, for the transverse efficiency w' of this bar due to tensile resistance

$$w' = \frac{4\,a'\,d'\,c}{l} = \frac{4 \times .64 \times 1.2 \times 4.5}{24} = .576 \text{ ton.}$$

But in rolled iron round beams the resistance of flexure has been found to be .7 that of tension, therefore for the total efficiency of the bar

$$w = .576 + (.576 \times .7) = .979 \text{ ton,}$$

which corroborates the result already given by the ordinary Formula.

Required the safe central load w for a solid round beam.

With the following exceptions the symbols have the same value as those given (**72**).

Let a' = the effective area of resistance = .204 a ⎫
d = diameter of the beam   ⎬ in inches.
d' = effective depth of section = .6 d ⎭
p = resistance of flexure = $1\frac{1}{2}\,c$ for cast iron.
,, = ,,   ,,   = .8 c for rolled steel.
,, = ,,   ,,   = .7 c for rolled iron.

$$\text{Then } w = \frac{4\,a'\,'d\,F}{l}.$$

CHAPTER VII.

GIRDER OF THREE MEMBERS.

75. It has been shown in the three preceding articles that the strength elements of a beam resulting from compressive and tensile resistance may be represented by the substitution for its section of two equal areas of equal resistance per unit of surface, one area representing the resistance to tension and the other the resistance to compression.

The difficulty, however, found with respect to solid beams, of bringing their theoretic efficiency into accordance with that shown by actual experiments on their transverse strength, led to the admission of another and very considerable strength element, the resistance of flexure.

This afforded a reason for the strength evinced by solid beams, of which no satisfactory explanation had been hitherto given.

Now the two equal areas of equal resistance by which the section of a solid beam has been replaced are typical of the actual top and bottom tables of a girder of three members. The resistance of flexure has, however, been found to be a comparatively small factor of strength in such girders, and in practice its value is covered by the ordinary formulæ given in the two following articles.

H

The determination of vertical or shearing stress at any vertical is a very simple matter, and has been fully exemplified throughout Chapter 2.

The effects of diagonal stress, in either the web of a cast-iron or in that of a plate girder, have for practical purposes been already disposed of (**22**).

It only remains therefore to add some useful formulæ for the strength elements of the top and bottom tables. These important members govern the efficiency of a girder; but owing to practical exigencies of construction the scantlings of the web will always exceed those ascertained by theory.

76. Formulæ for Central Loads.

The theory of the determination of the moments of horizontal stress in a beam (**24, 27**) will now be applied in the treatment of the strength elements of the top and bottom tables of a girder. Under any condition of loading, the stress in each of those members will be equal, but of a reverse nature, compressive in the top and tensile in the bottom table.

Let a = the area of either table in inches.

    d = the extreme depth   ⎞ *both* either in

    l = length between supports   ⎠ feet or inches.

    c = coefficient, Table 1 (**62**).

    w = safe load at centre of span.

The numerical factor 4 in each equation (21, 22) results from the resistance of each support or $\frac{w}{2}$ and the leverage at the centre of the span or $\frac{l}{2}$.

Given *a plate-iron girder* 20 feet long, 1.5 foot deep, with a load of 27 tons at its centre, to find the efficient area for each table at the centre of the span.

$$a = \frac{w\,l}{4\,d\,c} \qquad\qquad (21.)$$

$$= \frac{27 \times 20}{4 \times 1.5 \times 4.5} = 20 \text{ ins.}$$

Given the area of either table at the centre of the span to find the safe central load.

$$w = \frac{4\,a\,d\,c}{l} \qquad\qquad (22.)$$

$$= \frac{4 \times 20 \times 1.5 \times 4.5}{20} = 27 \text{ tons.}$$

77. Formulæ for Distributed Loads. As the centre of gravity of each half of the load is midway between the centre of the beam and a support, the effective leverage becomes $\frac{l}{4}$, and with w' as a distributed load the effect of the load on that part is $\frac{w'}{2}$, thus $4 \times 2 = 8$ the numerical factor. With this exception, the equations will remain as before.

For the area of either table at the centre of the span

$$a = \frac{w'\,l}{8\,d\,c} \qquad\qquad (23.)$$

Given the area of a table at the centre of the span, the safe load

$$w' = \frac{8\,a\,d\,c}{l} \qquad\qquad (24.)$$

Or, a = half the area of table required for an equal central load, and $w' = 2w$ or twice the weight of a safe central load.

It should be noted that for a cast-iron girder, $c = 1.4$ ton for the bottom table and 6 tons for the top table.

Having determined the sectional area of each table at the centre of the span, the relative proportional area at any other part of the span may be readily found (**23**) Fig. 9, and (**26**) Fig. 11. But these proportional areas require in practice to be modified to suit the nature of the work and material, and small girders are for simplicity frequently made of the same section throughout.

Should the depth of a girder be reduced from the centre of the span to each end, the theoretic area as determined for each table of a girder of uniform depth should at any given vertical, right and left of the centre, be increased inversely with the decrease of depth. Thus if the depth were reduced by one-half, that area should be doubled.

78. Strength Elements. Let A B C D, Fig. 35, be the transverse section of a cast-iron girder 12 feet long between the supports, 1 foot deep with tables 9 inches wide, and 1 inch metal throughout, then by Formula (22), (**76**)

$$w = \frac{4\,a\,d\,c}{l} = \frac{4 \times 9 \times 1 \times 1.4}{12} = 4.2.$$

therefore practically the safe load at the centre of the span is 4.2 tons.

To determine the efficiency of the strength elements in accordance with the treatment described (**72, 73**), draw

the diagonal dotted lines A D and B C crossing each other
in the neutral axis N N, at the centre of gravity of the
section. Assume the central load
to act in the direction of
the arrow P, then the half of
the section below the neutral axis
will be in tension, and the
hatched portions of it areas of
tensile resistance equal to that
existing at the bottom line C D
of the section. Now, **1.4** ton
being the coefficient of tensile
efficiency, the sum of the two
lower areas of equal resistance
each multiplied by its effective

Fig. 35.

leverage and by 1.4 will give the moment of the efficient
tensile strength of the section, thus for the

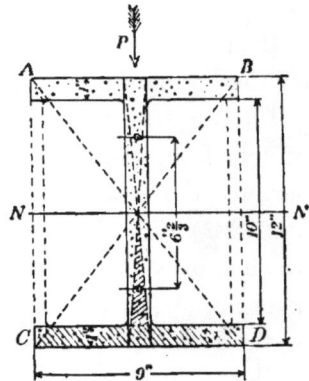

Table = $(8\frac{1}{4}'' \times 1'' =) 8\frac{1}{4}''$ area $\times 11''$ leverage $=$ 90.75.
Web = $(5 \times .417 =) 2.085$,, $\times 6\frac{3}{4}$,, $=$ 13.90.
                                                      104.65

Let m = moment of efficient tensile strength.
w = load at centre of girder.
l = length of girder between bearings in inches.
Then $m = 104.65 \times 1.4 = 146.51$

$$\text{and } w = \frac{4\,m}{l} = 4.07. \qquad (25.)$$

In order to determine the efficiency of the *resistance of
flexure* due to the section multiply the load w, provided
for by *tensile* resistance, by the coefficient p, which for cast
iron = $1\frac{1}{4}c$, (**72**), then 4.07 $\times$ $1\frac{1}{4}$ = 5.08.   And for a

flanged girder multiply the sum thus given by the thickness of the web and divide by the width of the tables.

Add the result to the value of w already obtained.*

$$\text{Then } \frac{5.08 \times 1}{9} = .56.$$

And w now $= 4.07 + .56 = 4.63$ tons.

It thus appears that the usual formula which gives 4.2 tons leaves a balance of one-tenth to the credit of strength.

79. *As an illustration of the error*† arising from a *rule* which is sometimes given for determining the strength of a flanged girder by the usual formula (12), (**67**), based upon experimental tests of the transverse strength of rectangular bars, assume A B C D, Fig. 35, to be a solid rectangular beam. Then assume the two spaces left between the rectangular outline and the section of the flanged girder to form together the section $8'' \times 10''$ of a second rectangular beam, and let each of these beams be 12 feet long between the supports. With Formula (12) determine the efficiency in cast iron of each of these solid beams, subtract the second from the first result, and the remainder will be the presumed efficiency of the flanged girder.

1. Thus $w = \dfrac{b\,d^2\,n}{l} = \dfrac{9 \times 12^2 \times .175}{12} = 18.9$

2. $\quad w = \dfrac{8 \times 10^2 \times .175}{12} = \underline{11.6}$

$$7.3 \text{ tons.}$$

Now 4.2 tons has been shown to be the proper safe central load for this girder, but the last result, 7.3 tons, gives an excess of three-fourths of that load.

* Experimentally demonstrated by Mr. Fairbairn, B. Baker on " Beams, Columns, and Arches."

† An error which occurs in some handbooks.

The same *rule* is wrong also for a box girder. For suppose the web to be split into two thicknesses, and a half thickness to be placed at each side, as shown by dotted lines in the figure, then all strength elements would remain precisely the same.

This error has arisen mainly from the non-recognition of the fact that a considerable proportion of the strength of a solid beam is due to the resistance of flexure. But there is also an error in assuming the stress at the top and bottom of the subtracted or inner beam when forming part of the entire solid beam to be the same as the corresponding stress in an independent beam, whereas in the example given it is only $\frac{10}{12}$ths of it, and the effect of this error is partially to counteract that of the other.

80. A Circular Hollow Beam is strictly of the same order in respect to strength elements as a girder of three members. A circular hollow *crane post* is a beam of this kind placed upright and resisting transverse lateral thrust. Let Fig. 36 be the cross-section of a hollow concentric cast-iron beam of 18 inches external and 15 inches internal diameter, and assume it to lie horizontally upon two supports placed 18 feet apart, and to be subjected to a load or thrust midway between the supports acting in

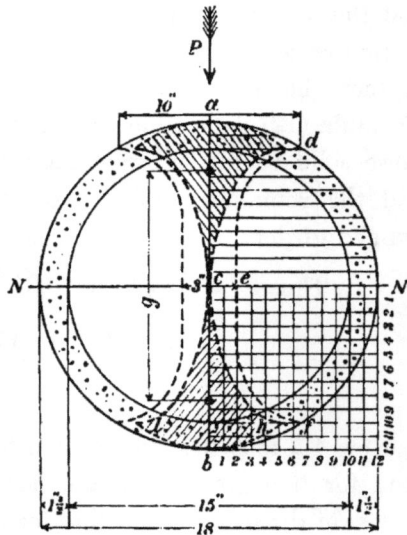

Fig. 36.

the direction of the arrow P. Then the horizontal line
N N passing transversely through the centre of the cylinder
will be the neutral axis of the section. Draw a vertical
centre line *a b* and any convenient number of horizontal
ordinates dividing the section of the beam into equal units of
depth. Point off from the centre line *a b* upon each ordinate
the thickness of the cylinder measured horizontally, and draw
through those points the curved dotted line *d e f*. Draw the
same curved line reversed upon the left side of the centre line *a b*.
Thus the annular section of the beam has been brought into
the form of a girder of three members with a top and bottom
table each 10 inches wide, and a web 3 inches thick at the
neutral axis of the section.

Draw vertical ordinates as shown in the figure, and
describe in the manner given (**73, 74**) the curved dotted
lines *c h b l c*. These will inclose the area of equal tensile
resistance, and the same curved lines relatively drawn in
the upper half of the section will also inclose an equal area
of equal compressive resistance per unit of surface.

Now in this example each of these areas = 22.5 inches;
while the vertical distance between their centres of gravity
= 1.07 feet, and this is the static depth *g*, Fig. 36, of the
beam. The effective safe strength *w'* of this beam due to
tensile resistance is, therefore, in the terms of a central load,
equation 22, (**76**).

$$w' = \frac{4\,a\,d\,c}{l} = \frac{4 \times 22.5 \times 1.07 \times 1.4}{18} = 7.49 \text{ tons.}$$

But for the total safe central load *w* to this result must be
added the resistance of flexure *p*, which should be determined

by the rule given (**78**) for a flanged girder, thus the actual width of the web being 3 inches and that of each table 10 inches.

$$p = \frac{7.49 \times 1\frac{1}{4}* \times 3}{10} = 2.80 \text{ tons.}$$

Then $w = 7.49 + 2.80 = 10.29$ tons = the safe central load.

Now the formula which has been sometimes given for determining the efficiency of a hollow cylindrical beam for a central load w when

$$D = \text{the outer diameter} \Big\} \text{ in inches.}$$
$$d = \text{the inner diameter}$$
$$n = \text{a transverse strength coefficient.}$$
$$l = \text{length of span in feet}$$

has been $w = \dfrac{D^4 - d^4\, n}{D\, l}$

which in the above example would be

$$w = \frac{18^4 - 15^4 \times .112}{18 \times 18} = 18.78 \text{ tons.}$$

Whereas the true safe load has been found to be 10.29 tons only, or little more than one-half of that given by the obsolete formula incorrectly based upon the transverse strength found by experimental tests to exist in *solid* beams.

The fourth power and the divisor $D\, l$ have been adopted in this equation because the stress at the surface of the inner cylinder d when in place has been assumed to be less than that at the outer surface of the large cylinder D in the proportion of their diameters.

* $p = 1\frac{1}{4}\, c$ for cast iron, page 90.

81. *The Surface of an Area*, such as one of equal static resistance, may be practically measured by drawing the curves defining the area full size or to a large scale, and then transferring the outline of the area on to a sheet of paper ruled, say, in $\frac{1}{8}$ or $\frac{1}{10}$ inch squares, by which means the required surface may be easily computed. Or this may be done with a piece of tracing paper so ruled and then applied over the drawing.

The Centre of Gravity of any such area or of the section of a girder can be found with considerable accuracy by drawing

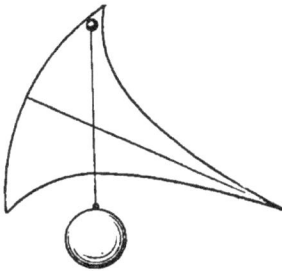

Fig. 37.

the outline of it upon and then cutting it out of cardboard or thin plate ruled with a centre line. Suspend the section thus prepared loosely upon a pin at a point in it as far removed from the centre line as may be convenient. Plumb down from the pin with a fine thread as shown, Fig. 37, and the centre of gravity of the section will be the point where the thread crosses the centre line.

The fourth power being the square of a square is easily found by a table of squares.

Thus, $13^4 = 13^2 \times 13^2 = 169^2 = 28561$.

The fourth root being the square root of a square root can be found in the same way by a table of square roots.

Thus for $\sqrt[4]{28561}$; $\sqrt{28561} = 169$ and $\sqrt{169} = 13$.

CHAPTER VIII

82. The Class of Pig Iron usually chosen for girders or other castings having when *in situ* to resist stress is No. 3 of the eight into which such iron is divided. No. 4 is, however, suitable for heavy girders. After a furnace has been tapped and the iron run into pigs, an external inspection is sufficient to enable an expert to identify and sort them into the eight classes.

When pigs are broken, No. 8 shows the whitest fracture, at No. 5 the fracture becomes mottled, and the lower numbers are various greys. These classes, however, merge into each other, and varieties are caused by temperature, and even by the state of the weather at the time of tapping the furnace.

Carbon is *mechanically* combined in pigs showing a grey fracture, and these are softer and tougher than those in which carbon being *chemically* combined consequently show a white and silvery fracture when broken.

Cast iron does not improve and may even deteriorate under successive remeltings, but a judicious mixture of scrap of other brands with pig iron facilitates melting, and when fused promotes fluency, the latter being a necessary condition of iron prepared for casting, which should quickly fill the moulds to every corner.

All mixtures suitable for girder work should on test show a fine evenly crystallised brightish grey fracture. But in the case of important series of castings, test bars should at intervals be cast simultaneously of the same metal. If the iron is of a good brand or mixture, bars 1 × 1 inch will when placed upon two supports stand up to a *pressure of* 800 *lbs.* applied at the centre of a clear span of three feet.

83. Board of Trade Regulations provide that cast-iron girders must not be used for bridges carrying lines of railway, "except in the form of arched ribbed girders where the material is in compression."

In a cast-iron arched bridge or in a cast-iron girder bridge carried over a railway, "the breaking weight of the girders should be not less than three times the permanent load due to the weight of the superstructure added to six times the greatest moving load that can be brought upon it."

"All castings for use in railway structures should, when practicable, be cast in a similar position to that which they are intended to occupy when fixed."

84. The Depth of Cast-iron Girders. Having already given (**76, 77**) formulæ for the strength and loading of girders generally, we may next consider certain conditions which regulate the proportions and efficiency of the leading members of a cast-iron girder.

The generally received rule with respect to the depth of such a girder is, that in order to ensure a reasonable amount of rigidity it need not be more than $\frac{1}{12}$ and should not be less than $\frac{1}{15}$ of the span. Collateral conditions or special circum-stances may, however, sometimes necessitate to some extent deviations from that rule. But in cast iron no advantage can be obtained by the adoption of an unusually deep web, while a girder of a reasonably compact section can be moulded and handled in the foundry with greater facility, and consequently at less cost.

With a depth determined thus solely in proportion to the length of the span, should the load be increased or doubled the area of each table must also be increased in a like direct proportion. This would, from the nature of all castings, involve also a thickening of the web, thus adding a perfectly useless amount of shearing resistance to that member, already by practical necessity theoretically much too thick, as will eventually be seen. Further, a girder thus proportioned might under a heavy load be less rigid, and would certainly require an useless amount of iron.

To judiciously meet these contingencies, the following formula (25*) is given for determining a suitable depth for a cast-iron girder. In this formula both length and load become factors.

Let $b\,w$ = the breaking weight

$w = \dfrac{b\,w}{5}$ = the safe load $\Big\}$ in tons at centre of span.

d = the depth of girder at centre of span $\Big\}$ in inches.
l = the length of span

Then $d = \sqrt[3]{\dfrac{5\,l\,b\,w}{7}}$ \qquad (25*.)

and $b\,w = \dfrac{7\,d^3}{5\,l}$ \qquad (26.)

85. For a Central Load, required a cast-iron girder of 40 feet clear span to carry 30 tons.

Now, when an open question, the first step will be to determine by Formula (25*) the depth d of the girder; and in this instance:

$$5\,l = 5 \times 40 \times 12 = 2400 \text{ inches,}$$
$$\text{and} \quad b\,w = 30 \times 5 = 150 \text{ tons.}$$
$$\text{Then } d = \sqrt[3]{\frac{2400 \times 150}{7}} = 37 \text{ inches.}$$

Thus having determined the depth, and the length of the span being $40 \times 12 = 480$ inches, the next step will be with an adaptation of Formula (21), (**76**), to resolve the moment of horizontal stress in the tables and then to determine the sectional area of each.

Let s = the moment of horizontal stress in each table.

a = the area of the bottom table.

a' = the area of the top table.

$$c = \begin{cases} 1.40 \text{ coefficient for bottom table.} \\ 6.00 \text{ coefficient for top table.} \end{cases} \text{Table 1, (62).}$$

$$\text{Then } s = \frac{w\,l}{4\,d} = \frac{30 \times 480}{4 \times 37} = 97.3 \text{ tons}$$

$$\text{and} \begin{cases} a = \dfrac{97.3}{1.4} = 70 \text{ inches for the bottom table.} \\ a' = \dfrac{97.3}{6} = 16.25 \text{ inches for the top table.} \end{cases}$$

The sectional area of each table having been found, it next becomes necessary to determine the width and thickness of each. Now, in order to avoid unequal shrinkage and to

ensure an efficiently homogeneous casting, it is essential
that the thickness of the web at top and bottom should
closely approximate to that of each relative table. Therefore
the thicker the tables the thicker the web. Thus the web even
in a well-proportioned cast-iron girder will of necessity be
of much greater shearing efficiency than is at all requisite,
and the tables cannot in practice be widened and thinned
down in order merely that the thickness of the web may be
reduced.

The matter of widths and thicknesses should therefore be
so settled as to impart sufficient lateral rigidity to the girder
and in accordance with practical exigencies, which may be
met in the following way.

The span and load of any cast iron girder being given,
and the depth d having then been determined by Formula
(25*), (**84**).

Let $\dfrac{3\,d}{4}$ = the width of the bottom table.

And let $\dfrac{3\,d}{10}$ = the width of the top table.

Then $\dfrac{d}{15}$ = $\begin{cases}\text{the mean thickness of the bottom table.} \\ \text{the thickness of the web at bottom.}\end{cases}$

„ $\dfrac{d}{25}$ = $\begin{cases}\text{the mean thickness of the top table.} \\ \text{the thickness of the web at top.}\end{cases}$

It is evident from Equation (26) that whatever the span
or load may be, the proportions above stated will remain

the same. Fig. 38 shows to scale the cross section of the required girder as determined.

The mean thickness of the tables is given because their thickness should taper from the web outwards, chiefly in order that the model or "pattern" of the girder may clear the sand in being lifted out of the mould.

In the example given, d being 37 inches,

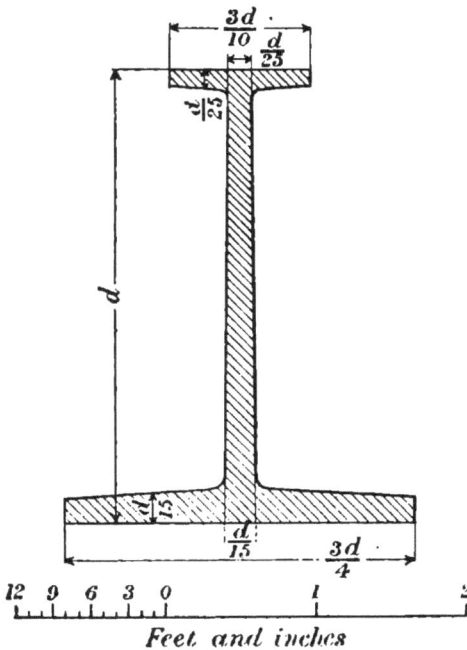

$$\frac{3\,d}{4} = 28 \text{ inches.}$$

$$\frac{3\,d}{10} = 11 \quad ,,$$

$$\frac{d}{15} = 2\tfrac{1}{2} \quad ,,$$

$$\frac{d}{25} = 1\tfrac{1}{2} \quad ,,$$

Feet and inches

Fig. 38.

bringing the results into practically measurable values when necessary by the addition of a small fraction.

For a girder 12 inches deep, the bottom table would be 9 inches wide by $1\frac{3}{10}$ inch thick, and the top table $3\frac{1}{2}$ inches wide by $\frac{1}{2}$ inch thick.

For a girder 48 inches deep, the bottom table would be 36 inches wide by $3\frac{5}{10}$ thick, and the top table $14\frac{1}{2}$ inches wide by $1\frac{13}{16}$ nch thick.

86. For a Distributed Load, first multiply the total load that the girder is intended to carry by 5, which will give the breaking weight, divide that by 2, and the result will be the equivalent b w at the centre of the span.

Then determine d, formula (25*), and the whole of the remaining scantlings for the central cross-section of the girder will follow in accordance with the proportions already given.

When cast iron is the material the theory of the stresses between the centre and ends of a girder carrying a distributed load (**26, 27**) cannot be even approximately followed as regards the thickness required for the web. When, however, circumstances do not require the depth to be uniform throughout, it may be advantageously reduced towards the ends, both as regards web and tables. Further a cast-iron girder should have, if possible, no lateral *ribs* at right angles with the web, for they are troublesome in moulding, somewhat interfere with continuity of section, and are not unfrequently the cause of unsoundness.

87. The Strength Elements of the girder given as an example (**85**) may also be determined in conformity with the method already explained.

Let A B D E, Fig. 39, be the cross-section of this girder. First determine the position of the *centre of gravity* C of this section through which point its *neutral axis must of necessity pass.* Square down lines from D and E, and draw the line d e parallel with D E, and at a vertical distance from C equal to that of the top A B of the girder from the same point. Complete the diagram with the eight lines radial from C, as shown in the figure.

Now the two hatched portions of the section within the

radial lines, the one above and the other below the neutral axis of the girder, will, if its position has been correctly determined, be two equal areas of equal action and reaction per unit of surface, viz. one of tensile and the other of compressive resistance. Each when computed will be found equal to 37 square inches.

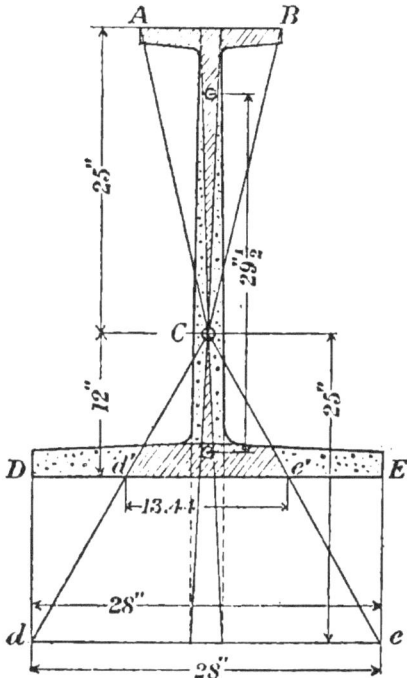

Fig. 39.

Having now these given areas, the next step in determining the efficiency of the section will be to find the value of the coefficient to be used. The tensile efficiency c, Table 1 (**62**), of cast iron is 1.4 ton per square inch, and this extends along the line from D to E at the bottom of the section. But the two areas being equal the coefficient for, or what is the same thing the assumed resistance of, each per square inch must be brought into the same relative terms.

Now as the efficiency of the actual section of the girder at the line D E is 1.4 ton per square inch, in order that the resistance of the portion d' e' of this line falling within the static area may represent the efficiency existing from D to E, the coefficient c must be increased in the ratio of D E to d' e', or in the proportion of the vertical dimensions 25″ and 12″,

$$\text{Thus } c = \frac{1.4 \times 25}{12} = 2.92 \text{ tons per square inch.}$$

This is also the resistance at the line A B, and consequently that, per square inch, of each of the static areas.

We must now add the resistance of flexure p. In this instance, the mean thickness of the web = 2 inches, and the mean width of the tables = 19½ inches. Therefore with these averages, and in conformity with the rule given (**78**) for this resistance,

$$p = \frac{2.92 \times 1\frac{1}{4}* \times 2}{19\frac{1}{2}} = .38 \text{ tons}$$

$$\text{and } c + p = 2.92 + .38 = 3.3 \text{ tons.}$$

The centre of resistance of each sectional static area being its centre of gravity, and those two points being in this instance vertically 29½ inches apart, that dimension becomes the effective static depth of the section, Fig. 39.

Having now brought the strength elements into relatively proportionate factors, let w be the safe central load, a the static area, and d the static depth,

$$\text{Then } w = \frac{4\ a\ d\,(c+p)}{l} = \frac{4 \times 37 \times 29\frac{1}{2} \times 3.3}{480} = 30 \text{ tons.}$$

The result of this investigation will therefore be found to corroborate that which would be given simply by the usual formula (22). At the same time it demonstrates, first, that

* $p = 1\frac{1}{4}\ c$ for cast iron, page 90.

that formula is substantially correct, although the *whole* depth of the beam is a factor; and secondly, that the provision made for compressive resistance is theoretically twice that absolutely required, because the maximum compression at the line A B is 2.92 tons only per square inch, instead of the safe allowance of 6 tons.

But practically it would be inadmissible to reduce the thickness of the web at top and also the top table to $\frac{3}{4}$ inch instead of $1\frac{1}{2}$ inch when lateral rigidity, shrinkage in casting, and the possibilities of accidental damage are taken into consideration. This will be still more evident in a girder 12 inches deep, the top table of which would by the rules laid down (**85**) be $\frac{1}{2}$ inch thick, and would consequently be left $\frac{1}{4}$ inch thick only, if similarly reduced.

CHAPTER IX.

GIRDER work is frequently made simply a matter of *contract* per ton for the whole amount required, ordinary and extra scantlings and weights inclusive.

In designing such work, it may, however, be useful to note the following particulars.

Iron Plates and Bars, Price Regulations.

88. Rolled Iron Plates not exceeding any one of the following conditions are included within ordinary current prices.

> 4 cwt. per plate.
> 30 superficial feet.
> 15 feet long.
> 4 feet down to 1 foot wide.

Extras on Iron Plates.

From 4 to 5 cwt. per plate	10/- per ton.	
,, 5 ,, 6 ,,	,,	25/- ,,
,, 6 ,, 7 ,,	,,	50/- ,,
,, 7 ,, 8 ,,	,,	75/- ,,
,, 8 ,, 9 ,,	,,	110/- ,,
Plates from 15 to 18 feet long	20/- ,,	
,, 18 ,, 25 ,,	40/- ,,	
,, 25 ,, 30 ,,	60/- ,,	

For every 6 inches or part of 6 inches in width exceeding 4 feet 20/- per ton.

89. └ **and** ⊤ **Iron** fall within current prices when their sections do not exceed 4 × 4 inches, or 8 inches run of section measured in each case from out to out.

From 8 to 10 inches run of section add 6 per cent. per ton to the current prices.

Up to 8 inches run of section the ordinary rolled lengths are 40 feet, and above 8 and up to 10 inches run of section the ordinary rolled lengths are 35 feet.

For every foot or part of a foot above the ordinary lengths add 2*s.* 6*d.* extra per ton.

The above conditions remain the same under any fluctuations of the iron market.

Abstract of Admiralty Tests for Rolled Iron.

90. Plate Iron. B " Best," B B " Best Best."

Tensile Tests.

B plates 20 tons with the grain without fracture.
,, 17 ,, across ,, ,,
B B plates 21 ,, with ,, ,,
,, 18 ,, across ,, ,,

Forge Test, cold. Plates to admit of being bent cold to the following angles without fracture.

B plates with the grain to an angle of 75 degrees.
,, across ,, ,, 30 ,,
B B plates with ,, ,, 90 ,,
,, across ,, ,, 40 ,,

Forge Test, hot.

B plates with the grain to angle of			90	degrees
,, across ,,		,,	60	,,
B B plates with	,,	,,	125	,,
,, across ,,		,,	90	,,

---

91. Bar Iron, rectangular and round.

Tensile Test lengthways 22 tons per square inch.

Forge Test, hot.—Bars are punched with a punch one-third the width or diameter of the bar at distances of one and a half and three widths or diameters of the bar from its end, the holes being at right angles to each other. The holes are then drifted out to one and a quarter times the width or diameter of the bar. The sides of the holes are then split and the ends must admit of being turned back without fracture. Fig. 40.

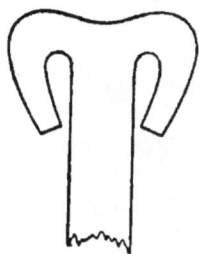

Fig. 40.

Forge Test, cold.—Bars are notched and bent as shown, Fig. 41, and the iron must then present a fibrous fracture free from cinder, &c.

Fig. 41.

92. L **Iron.**

Tensile Test lengthways 22 tons per square inch.

Forge Tests, hot.—Angle iron is bent as regards section in two reverse directions A and B, Fig. 42; it is also flattened

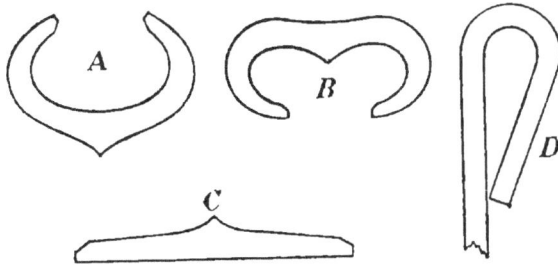

Fig. 42.

as C, and the end then bent over as D.

Forge Test, cold.—A sample is notched and broken to show the quality of the iron.

93. T **Iron.**

Tensile Test lengthways 22 tons per square inch.

Forge Test, hot.—The flange is bent in reverse directions E and F, Fig. 43, and the stem is also bent as at G.

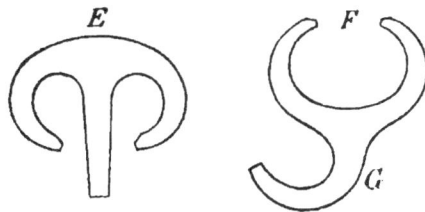

Fig. 43.

Forge Test, cold, is the same as for angle iron.

94. Board of Trade Regulations. In a wrought-iron or steel bridge the greatest load which can be brought upon it, added to the weight of the superstructure, should not produce a greater stress on any part of the material than five tons where wrought iron is used, or six tons and a half where steel is employed, per square inch.

Steel Plates and Bars.

95. Steel Plates are readily obtained when not exceeding in any particular the following table of ordinary thicknesses and dimensions.

TABLE 5.

$\frac{1}{16}$ inch thick,	28 feet super,		14 feet long,		4 feet wide.		
$\frac{3}{32}$	,,	31	,,	18	,,	$4\frac{1}{2}$	,,
$\frac{1}{8}$	,,	40	,,	22	,,	5	,,
$\frac{5}{32}$	,,	50	,,	25	,,	$5\frac{1}{4}$	,,
$\frac{3}{16}$	,,	65	,,	30	,,	$5\frac{1}{2}$	,,
$\frac{1}{4}$	,,	72	,,	33	,,	6	,,
$\frac{5}{16}$	,,	85	,,	35	,,	$6\frac{1}{4}$	,,
$\frac{3}{8}$	,,	98	,,	38	,,	$6\frac{1}{2}$	,,
$\frac{7}{16}$	,,	105	,,	40	,,	7	,,
$\frac{1}{2}$	,,	115	,,	40	,,	$7\frac{1}{2}$	,,
$\frac{5}{8}$	,,	105	,,	40	,,	$8\frac{1}{4}$	,,
$\frac{3}{4}$	,,	125	,,	37	,,	$8\frac{3}{4}$	,,
$\frac{7}{8}$	,,	125	,,	34	,,	$8\frac{3}{4}$	,,
1	,,	125	,,	31	,,	$8\frac{3}{4}$	,,
$1\frac{1}{8}$	,,	110	,,	28	,,	$8\frac{3}{4}$	,,
$1\frac{1}{4}$	,,	110	,,	25	,,	$8\frac{3}{4}$	,,

L *and* T *Steel Bars* are usually rolled to lengths of 35 feet.

Abstract of Admiralty Tests for Steel.

96. *Tensile Test.*—A piece of a bar, or a strip cut lengthways to have an ultimate tensile strength of not less than 26 and not exceeding 30 tons* per square inch of section, with an elongation of .2 inch in a length of 8 inches, or $2\frac{1}{2}$ per cent.

Bending Test.—A piece of a bar, or a strip cut crossways or lengthways $1\frac{1}{2}$ inch wide, heated uniformly to a low cherry red, and cooled in water of 82° Fahrenheit, must stand bending in a press to a curve of which the inner radius is one and a half times the diameter of the bar or thickness of the strip.

The Ductility of every plate or bar is ascertained by the application of one or both of the above tests to the shearings, or by bending them cold with a hammer.

With respect to percentage of *carbon*, steel is midway between cast and wrought iron, thus :

Cast iron contains from 6 to 2 per cent. of carbon.
Steel   ,,   ,,   2 to $\frac{1}{3}$,,   ,,
Wrought iron   ,, $\frac{1}{3}$ to $\frac{1}{4}$,,   ,,

The most efficient qualities of steel contain $1\frac{1}{4}$ per cent. of carbon, about which point its tenacity is greatest, and its ultimate elongation about $2\frac{1}{2}$ per cent.

---

* Because beyond this steel is liable to be less ductile and reliable.

97. Screwed Bolts and Rods, diameters of efficiency at the bottom of Whitworth threads.

TABLE 6.

Diameters of Bolts, inches.	Diameters less Threads, inches.	Diameters of Bolts, inches.	Diameters less Threads, inches.	Diameters of Bolts, inches.	Diameters less Threads, inches.
$\frac{1}{8}$	.093	$\frac{3}{4}$	.622	$1\frac{3}{4}$	1.494
$\frac{3}{16}$	.134	$1\frac{13}{16}$	.684	$1\frac{7}{8}$	1.590
$\frac{1}{4}$	.186	$\frac{7}{8}$	.733	2	1.715
$\frac{5}{16}$	.241	$1\frac{15}{16}$	.795	$2\frac{1}{8}$	1.840
$\frac{3}{8}$	.295	1	.840	$2\frac{1}{4}$	1.930
$\frac{7}{16}$	.346	$1\frac{1}{8}$	.942	$2\frac{3}{8}$	2.055
$\frac{1}{2}$	.393	$1\frac{1}{4}$	1.067	$2\frac{1}{2}$	2.180
$\frac{9}{16}$	.456	$1\frac{3}{8}$	1.161	$2\frac{5}{8}$	2.305
$\frac{5}{8}$	.508	$1\frac{1}{2}$	1.286	$2\frac{3}{4}$	2.384
$\frac{11}{16}$	.571	$1\frac{5}{8}$	1.369	3	2.634

The width across the flats ranges for *six-sided nuts,* from the smallest to the largest bolt given in the table, from twice to one and a half times the diameter.

The usual practical depth for a nut is the diameter of the bolt, although theoretically a much less depth would render the stripping resistance of the thread equal to the tensile resistance of a bolt.

Take for instance a $1\frac{1}{4}''$ bolt, assume that its efficiency is reduced by the thread to that of a bar $1''$ diameter, and that the tensile and shearing efficiencies of the material are equal.

Then a collar $\frac{1}{4}$ inch deep upon the reduced bolt would present a shearing resistance equal to its effective tensile resistance, because the circumference multiplied by $\frac{1}{4}$ = $\frac{3.1416}{4}$ = .7854 or the area of the bolt, less thread.

But assuming the shearing efficiency of a nut to be reduced one-half by the thread, the theoretical depth of the nut for a $1\frac{1}{4}$-inch bolt would then be $\frac{1}{4} \times 2 = \frac{1}{2}$ inch, or only $\frac{2}{5}$ of $1\frac{1}{4}$, or of what would be the usual depth. When, however, nuts are liable to much wear, as, for instance, in workshop tools, their depth should be further proportionally increased.

Again, although the usual depth of a nut is the diameter of the bolt, it may, under certain conditions, and both as regards weight and appearance, be advantageously much less.

A reduction of depth or thickness may, for instance, be made when a nut is adopted merely for the purpose of securing the members of a system together, and not for taking any direct stress. Such is the duty of nuts at the ends of bolts connecting the links of suspension-bridge chains.

In *long tie-bars or rods* weight may be reduced by shutting for instanc $1\frac{1}{4}$-inch screwed ends on to a 1-inch bar, when the whole will have the efficiency of a bar 1 inch in diameter. For its tensile efficiency would evidently be no more if with screwed ends it were $1\frac{1}{4}$ inch in diameter throughout.

98. Riveting and Rivets. In order to insure thoroughly efficient work, the shanks of rivets should always be so extended laterally by pressure in the act of riveting as not only to completely but tightly fill the rivet holes. Thus they will all take a bearing when in place, and each rivet will be in a condition to effectually take its part in resisting lateral compressive stress.

Now machine as com-
pared with hand riveting
is not only more rapidly
performed and less costly,
but it is at the same
time a more perfect ope-
ration. For the steady
hydraulic riveting pres-
sure given at one stroke
of the machine passes

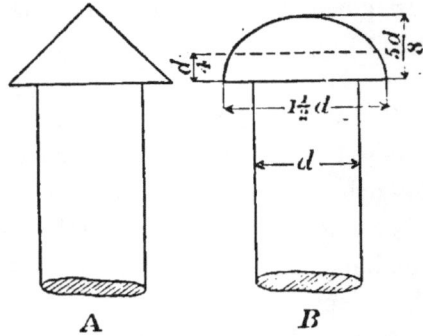

Fig. 44.

simultaneously throughout the whole body of a rivet, compel-
ling it to fill the rivet hole far more effectually than can
possibly be the result of the successive blows of hand-riveting.
Workmen, however, were very proud of what they deemed
the beautifully finished conical rivet heads of handwork,
terminating in a finely-formed point, as Fig. 44, A. This
high finish has been found, however, to be worse than useless, for the elaborate process of the often-repeated and comparatively light blows of a hand hammer necessary to produce
it crystallises the interior of iron rivets, sometimes rendering
them so brittle that their heads have been known to start
from their shanks. Consequently the process of hand-
hammer riveting must considerably reduce the efficiency of
the work.

That repeated blows from a hammer may render steel also
brittle becomes evident from an accidental vertical fracture
of the upper end of an old cold chisel. For such an occur-
rence will show that coarse, bright crystallisation of a central
portion of the interior has taken place in the form of a central

acute-angled cone with its base at the hammered end of the chisel.

These facts have led to the general adoption of the " snap " or cup head for rivets, which in hand work is finished and in machine work entirely formed with a cup-shaped tool. This is the best form both for the head and tail of rivets for machine work. It besides has certainly a neater appearance in girder work than the conical form of head.

The mean *tensile strength* of rivet iron is about *26 tons per square inch.* The diameter of a cup head should be $1\frac{1}{2}$, and the height of it $\frac{4}{8}$ the diameter of the rivet. For the former will give a bearing surface on the plate fully equal to the sectional area of the rivet, and the latter will provide a practically better formed head, and one not so rapidly chilled as a cone during the process of riveting. Theoretically the effective stripping or shearing height of the head would be one-quarter the diameter of the rivet, as shown by the dotted line, Fig. 44, B, because, as stated in the last Article, the circumference of a cylinder multiplied by $\frac{1}{4}$th its diameter equals its area,

$$\text{or } \frac{3.1416}{4} = .7854.$$

99. Plates and Rivets. The following table gives the diameter of rivets as usually adopted, either in iron or steel girder work, for various thicknesses of plate, also in each case the sectional area of the rivet and the area of its effective lateral bearing surface in the plate; the latter being the diameter of the rivet multiplied by the thickness of the plate.

TABLE 7.

| Plate. | Rivets. | | Effective Lateral Bearing. |
Thickness.	Diameter.	Area.	
inch.	inch.	square inch.	square inch.
$\frac{3}{16}$	$\frac{1}{2}$	.196	.094
$\frac{1}{4}$	$\frac{5}{8}$	.306	.156
$\frac{5}{16}$	$\frac{5}{8}$	.306	.195
$\frac{3}{8}$	$\frac{3}{4}$	.441	.281
$\frac{7}{16}$	$\frac{3}{4}$	.441	.328
$\frac{1}{2}$	$\frac{3}{4}$	.441	.375
$\frac{9}{16}$	$\frac{7}{8}$	.601	.492
$\frac{5}{8}$	$\frac{7}{8}$	.601	.547
$\frac{11}{16}$	$\frac{7}{8}$	.601	.601
$\frac{3}{4}$	1	.785	.750
$\frac{13}{16}$	1	.785	.812
$\frac{7}{8}$	1	.785	.875
1	$1\frac{1}{8}$	.994	1.125

Apart from theoretical requirements, when corrosion and painting are taken into consideration, it is not desirable in outdoor work to use iron plates of a thickness less than $\frac{1}{4}$ inch, nor steel plates less than $\frac{3}{16}$ inch thick.

To avoid extra cost arising from the use of large-sized or heavy plates, it is frequently desirable to select thicknesses suited to the nature of the proposed work, and then to build up the total thickness required in layers of such thicknesses, as for instance is often done in the tables of girders.

The side and end edges of all plates should be planed so that no irregularities in those parts should be left to attract or harbour moisture. Besides, by that stipulation, defects might become apparent, good butt joints could be made, and, after painting, work thus completed has a far neater appearance.

In riveting together a series of thicknesses of plate, as may be the case for the tables of a girder, if the diameter of the rivets is fixed by the thickest plate or angle iron they have to pass through, it need not exceed the allowance given in Table 7. For such work rivets are just as efficient in passing through several as through two thicknesses of plate only.

Drilling rivet holes insures far better work than punching. Rivet holes in steel plates should always be drilled, as steel is more fatigued and injured by punching than rolled iron. Manufacturers provided with special machinery for the purpose would as soon drill as punch rivet holes.

Steel rivets should be of the best refined mild soft steel.

100. The Pitch of Riveting, or the linear distance apart of rivets from centre to centre, may be varied with the loading and with the consequent constructive requirements of a girder. A longitudinal pitch of 4 inches is, however, usually found sufficient, and adopted for the tables of girders, without reference to the number of thicknesses of plate through which the rivets have to pass. With thick, say 1-inch, plates, however, and a very heavy load per foot run, a 3-inch longitudinal pitch may become necessary; but any way, the *transverse* distance apart of these rivets must be adjusted to suit the width of the tables and that of the angle-irons connecting them with the web plate. It may also sometimes be advisable to adopt a 3-inch pitch for the rivets of the vertical joints in a web in order that they may provide an area of bearing surface sufficient to resist vertical stress.

It is evident, therefore, that in a series of rivets extending over a given distance, their bearing surface and also their shearing area increases inversely with decrease of pitch, also that each may be doubled or further increased by the adoption of two or more rows of rivets on each side of a vertical joint in the web plates.

101. Joints and Covers. The meeting endways or sideways of two plates with each other in order to form a line of plates is termed a joint.

In girder work a line of plate when thus broken is necessarily united at each such joint with a cover. A joint may be covered with a single piece of plate or with a plate riveted on each side of the plates to be united, according to the requirements of the work. Joints in web plates are frequently covered on each side by the necessary vertical T iron stiffeners, which thus serve two purposes.

Plates and covers when in tension thus form as it were the links of a chain. But covers are also necessary when a line of plates is in compression, because it is practically impossible in building up work to bring plates into perfectly efficient abutting contact with each other. Besides, the plates forming either table have in imparting lateral rigidity to a girder a tendency to oscillate or work laterally at the joints, and that tendency is resisted by the covers.

Required efficiently to unite longitudinally two separate plates A C and C B, Fig. 45, each $7\frac{1}{2}$ inches wide and $\frac{1}{2}$ inch thick, with one cover of the same width and thickness.

Let two lines of $\frac{3}{4}$-inch rivets be adopted passing through

K

the plates and cover, and assume the efficiency of the
material to be the same for all strains. Then as the lateral

Side of Plates

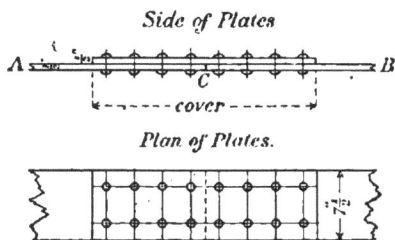

Plan of Plates.

Fig. 45.

bearing efficiency of a
$\frac{3}{4}$-inch rivet in a $\frac{1}{2}$-inch
plate is less than the
shearing efficiency of the
rivet, Table 7 (**99**), in
order efficiently to cover the
joint the lateral bearing
area of the rivets on each
side of it should equal the
efficient area of each plate, which for the present purpose
may be taken as simply its width, less rivet holes, or

$$7\tfrac{1}{2} - (\tfrac{3}{4} \times 2) = 6.$$

In the same way, the lateral bearing efficiency of a rivet
may be taken as its diameter, the factor of depth or thickness being common to both plate and rivet.

Thus as the efficiency of the plate is 6, and that of a rivet
is $\frac{3}{4}$, the number of rivets required on each side of the joint

$$= \frac{6}{.75} = 8.$$

Or, the width of plate less rivet holes divided by the
diameter of rivet equals the number of rivets required, up to
a $\frac{3}{4}$-inch plate with 1-inch rivets, beyond which as the lateral
bearing then becomes greater than the sectional area of the
rivet, the net sectional area of the plate must be divided by
the area of the rivet, and the product will be the number of
rivets required on each side of the joint.   See Table 7 (**99**).

In the case of the table of a girder formed of several tiers
of plates, work may be simplified by bringing necessary joints

under one cover plate when they can conveniently be so arranged.

Given the table of a girder $17\frac{1}{2}$ inches wide, formed of four thicknesses of $\frac{9}{16}$-inch plate riveted together with four longitudinal lines of $\frac{7}{8}$-inch rivets, Fig. 46, to unite four joints or one in each line of plate under one cover plate $\frac{9}{16}$ thick.

Now the efficiency of each plate represented lineally $= 17\frac{1}{2} - (\frac{7}{8} \times 4) = 14$, and in the same way that of each transverse line of rivets $= \frac{7}{8} \times 4 = 3\frac{1}{2}$. Then $\dfrac{14}{3\frac{1}{2}} = 4$ transverse lines, of four rivets each, required on each side of each joint as shown in the Figure.

Side of Table.

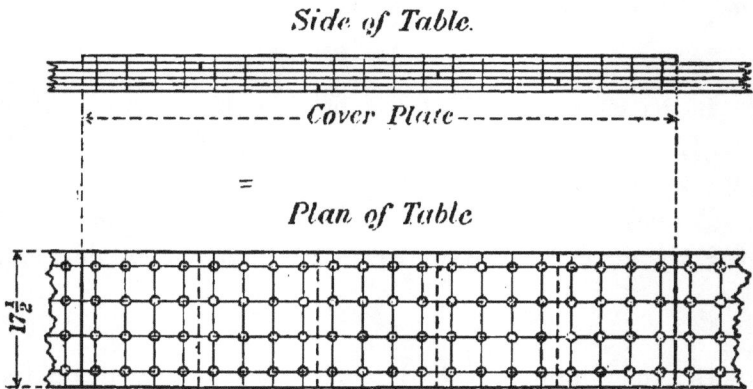

Fig. 46.

In the case of detached tie-bars or plates, the best arrangement would be that of a cover plate on each side of the joint, the two plates giving together the total thickness required. This arrangement not only avoids the cantilever action of the rivets of the ordinary "lap joint," but gives a double shearing efficiency to each rivet.

The most simple covering for a joint in a line of angle iron connecting the web with a table of a girder is a plain piece of plate of suitable cross-section, secured as may be most convenient to the upper or under side of the table by the ordinary riveting.

It will be obvious that in ordinary work the most practical arrangement for riveting is in longitudinal and transverse straight lines, the latter being set out at right angles with the former, as shown by the lower diagram, Figs. 45 and 46.

102. The Proportionate Cost per Ton at the works, of girders built up of plates and ⌐ and T bars may be taken as 10.25 for iron and 11.25 for steel, whatever may be the price of iron in the market.

Now by Table 1 **(62)**, the efficient strength for all stresses may be taken as :—

$4\frac{1}{2}$ tons per square inch for rolled iron.
$6\frac{1}{2}$,,   ,,   for rolled steel.

Then for the proportionate cost of an efficiency in iron of $6\frac{1}{2}$ tons,

$$\frac{10.25 \times 6.5}{4.5} = 14.8$$

But in steel, the same efficiency costs in proportion 11.25, and therefore the cost of iron compared with that of steel is as 14.8 to 11.25, or there is a gain of 24 per cent. in the use of steel.

Notwithstanding, however, the greater ultimate strength of steel, rolled iron is more ductile, tough, and pliable, and iron plates possess great evenness of character. For light work therefore, especially if liable to sudden shocks and strains, iron may be used with advantage, while at the same time under such circumstances greater weight and consequent inertia would be an item in its favour.

103. **The Depth of Plate Girders** when not of necessity determined by special circumstances such as limited space in height or of headway is as has been shown in the case of cast-iron girders a matter of primary consideration.

For with a given span and load, the depth of a girder when determined governs not only the relative proportions of the subsequent scantlings, but also to some extent the amount of material required for its construction.

From the special adaptability of rolled plates and bars to the building up of girders, the requisite dimensions and thicknesses of their members can be satisfactorily carried out without any difficulty. Thus more advanced and economic forms of section can be adopted than are in any way possible in cast iron.

Now the *effective* vertical stress of the load at a support multiplied by half the span of the web measured in units of its depth will always be equal to the total vertical stress upon the horizontal section of the half web, and to the horizontal stress upon the central section of each table. This is so with either a central or with an evenly distributed load (**23, 27**).

Thus with the same span and load, and assuming the re-sistant strength of the material employed to be the same for all stresses, with double the depth of web, half the sectional area would be required in each table, and likewise half the horizontal sectional area in the web plate. Half the weight of metal would thus be saved in the tables; but as the web has been doubled in depth, its weight would remain the same.

Theoretically, therefore, the weight of the web is a constant quantity, while it would seem that that of the tables could be reduced to any extent; but practically this is not so, for a plate web should never be less than ¼ of an inch thick, while increased depth combined with reduction of thickness would reduce its lateral rigidity, and so necessitate an increase of strength in the usual vertical stiffeners.

Besides, vertical stress caused by an evenly distributed load may in the case of a rolling load of the same weight per foot run be considerably increased in the central portions of the web (**40**). Other matters of detail may also vary the amount of materials required both in the web and tables, such as stiffeners, cover plates, and rivet heads.

Now it is evident that the best ratio of depth to length for any girder is that which provides the greatest efficiency with the least amount of material. This is, however, not an easy matter to determine. It cannot be settled by direct theory, but may be approximated by the co-ordination of many contingent conditions of load, span, and structure, but especially of the former.

The results of comparatively recent experience both at home and abroad have, however, demonstrated that it has been found advantageous to give greater depth to plate girders than had been previously usual.

Sir Benjamin Baker, in closing his valuable review of this subject, observes, that he has not given " a general expression applicable to any given case ; because, even if it were possible to do so, the complexity of the formulæ would rather conceal than illustrate the relative importance of the different conditions affecting the case." He continues, " we have, therefore, preferred completing the results of certain specific cases and arranging them graphically in the form of a diagram.*

* " The Strength of Beams, Columns, and Arches." B. Baker, 1870

It is from that diagram that the following, Table 8, of depths for plate girders has been computed, extending from 20 to 200 feet span, and from a load of 10 to 100 cwt. per foot run.

TABLE 8.

DEPTHS FOR PLATE GIRDERS.

Feet span. Per ft. run. cwt.	20		40		60		80		100		120		140		160		180		200	
	Ft.	ins.	Ft.	ins.	Ft.	ins.	Ft.	ins.	Ft.	ins.	Ft.	ins.	Ft.	ins.	Ft.	ins.	Ft.	ins.	Ft.	ins.
10	1	8	3	3	4	9	6	3	7	8	9	1	10	5	11	9	13	0	14	3
20	2	1	4	2	6	2	8	1	9	11	11	0	13	5	15	2	16	9	18	4
30	2	6	4	11	7	4	9	8	11	11	14	1	16	3	18	5	20	6	22	5
40	2	9	5	6	8	0	10	8	13	2	15	8	18	2	20	7	22	11	25	4
50	3	0	5	11	8	8	11	6	14	3	17	0	19	10	22	3	24	10	27	5
60	3	2	6	3	9	3	12	3	15	2	18	0	20	10	23	7	26	4	29	0
70	3	4	6	6	9	9	12	10	15	10	18	11	21	10	24	9	27	7	30	4
80	3	5	6	9	10	1	13	4	16	6	19	9	22	9	25	10	28	9	31	8
90	3	6	7	0	10	5	13	10	17	1	20	4	23	6	26	8	29	9	32	10
100	3	7	7	2	10	8	14	1	17	5	20	8	23	11	27	2	30	3	33	4

Mr. Fairbairn stated at the Institution of Civil Engineers in 1850, that the depth of girders above 150 feet span should not exceed $\frac{1}{15}$th of the span, in order that by a low centre of gravity oscillation caused by a passing load might be reduced. Below that span, he recommended a depth of $\frac{1}{13}$th as being more economical; but Sir Benjamin Baker in allusion to such limitations observes, " At the present time (1870) the depths of such girders would probably be made one-half greater."

Without attaching undue importance to minor fractions, Table 8 may be taken as affording ready data for reference in the matter of depth.

It will be observed that the ratio of the depth somewhat decreases with the span ; thus with a load of 10 cwt. per foot and a 20-feet span, the depth given is 20 inches, or $\frac{1}{12}$th, while with a span of 80 feet the depth is 75 inches, or $\frac{1}{12.8}$th. But the depth materially increases with every increase of load.

104. Constructive Details of a Plate Girder

to carry a load of 100 tons equally distributed over a clear span of 40 feet.*

With a given amount of material the most efficient longitudinal form for the web of a plate girder is a rectangular parallelogram. For when the central depth of the girder has been determined, any reduction of that depth extending endways from the centre would cause no alteration in the weight of the web, because its shearing efficiency at any given vertical should remain constant. But the sectional area and consequently the weight of the tables should be increased in an inverse proportion to any reduction of web depth, and moreover, any variation of that depth would render the construction of the girder less simple throughout.

Now as the depth or thickness of the tables is, excepting for small plate girders, usually decreased from the centre to the ends, so when the depth of the web remains constant it is for convenience usually taken as the depth of the girder. Thus

* 40 feet is, except for important thoroughfares, the usual width of new streets, being the minimum allowed by a bye-law usually passed by the Local Authority in conformity with that made by the Metropolitan Board of Works under the Metropolis Local Management Act and adopted by the London County Council.

the top and bottom of a rectangular web are two parallel straight lines, either of which may be used as a datum for calculations or construction. Figs. 47 and 48 (next page).

105. Table Details. The given span being 40 feet or 480 inches, and the load 100 tons distributed, or 50 cwt. per foot run, referring to Table 8 (**103**) let the depth of the girder be 5ft. 10 inches = 70 inches.

Then for the sectional area of each table at the centre of the span adopting steel as the material to be used,

By Equation (23), (**77**),

$$a = \frac{w\,l}{8\,d\,c} = \frac{100 \times 480}{8 \times 70 \times 6\frac{1}{2}} = 13.2 \text{ inches.}$$

This area will be provided partly by plates and partly by angle bars as next explained.

Let each table be connected with the web by two $3\frac{1}{2}'' \times 3'' \times \frac{3}{8}''$ angle bars, and let $\frac{3}{4}''$ rivets be used throughout the girder with a 4-inch longitudinal pitch for the tables.

Taking the section of the two outer $3\frac{1}{2}''$ flanges of these angle bars, less two rivet holes, as parts of the total area required for each table.

$$(2 \times 3\frac{1}{2}) - (2 \times \frac{3}{4}) = 5\frac{1}{2} \text{ and } 5\frac{1}{2} \times \frac{3}{8} = 2.$$

Then $13.2 - 2 = 11.2$ inches or the sectional area to be provided by the table plates.

The determination of the width of the tables must to some extent be the empiric result of a practical balance of contingent conditions. As the bottom table is in tension, and thus not liable to distortion in resisting that strain, its width is theoretically so far of little importance. That, however, is not so with the top table, which, as in the case of a column, may be laterally deflected by longitudinal compression. But for simplicity of construction, the same width is best adopted

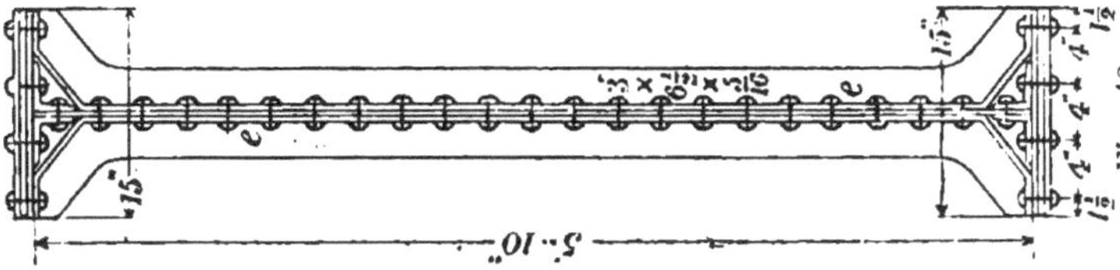

Fig. 48.

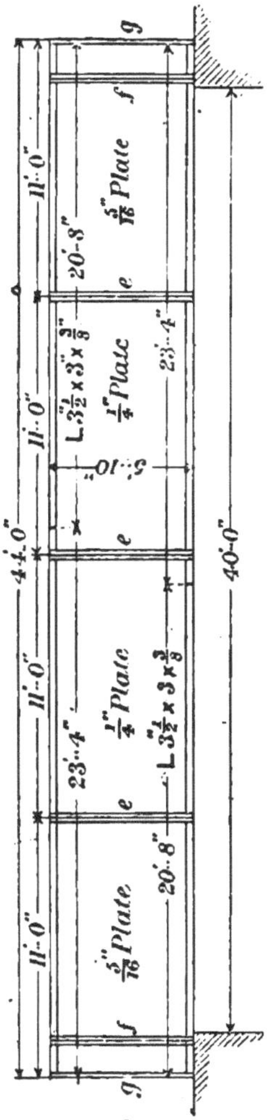

Fig. 47.

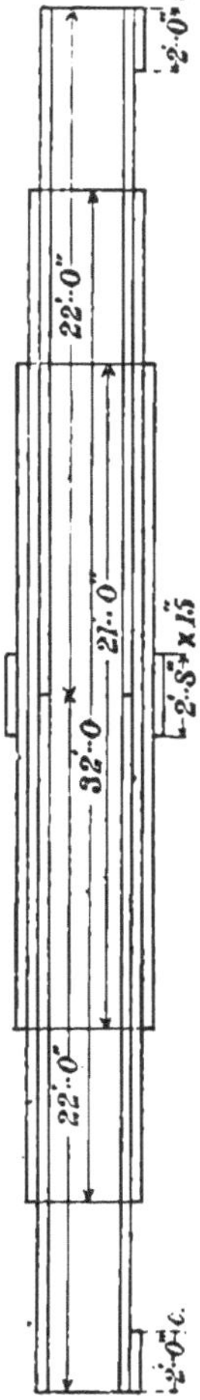

Fig. 49.

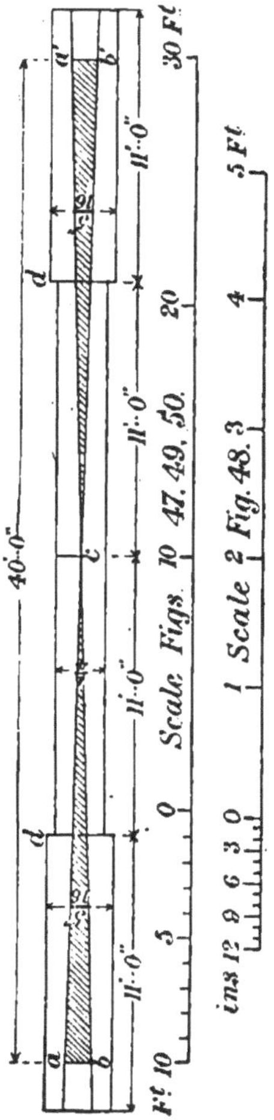

Fig. 50.

for each table. Besides the two tables serve together as stiffeners to the web and thus to the girder against the stress of wind or any other cause of lateral disturbance.

Assuming the width of each table to be 15 inches, and deducting the width of four $\frac{3}{4}''$ rivet holes, or $15 - (\frac{3}{4} \times 4)$ $= 12$ inches, or the effective width of the table, we then have $\frac{11.2}{12} = \frac{15}{16}$ or a thickness of three $\frac{5''}{16}$ plates, will give the sectional area required for the central part of each table.

Then, bearing in mind the fraction of the whole area required for each table supplied by the outer flange of the angle bars, the *lengths* of the table plates may be arranged as shown in diagram Fig. 49, so as to provide the proportional area required at every part of the span for resisting the horizontal stress set up by a distributed load (**27**).

Let the inner line of plate of each table be in two 22-feet lengths connected by a cover plate.

Now as the net transverse run of section of each plate less rivet holes is 12 inches, and the coefficient being the same for all resistances, the aggregate run of rivets in the cover plate should also be 12 inches on each side of the joint (**101**), and as the rivets are $\frac{3}{4}''$ diameter the number of rivets required $= \frac{12 \times 4}{3} = 16$, or there should be four rows of four rivets each on each side of the joint. This with a 4-inch longitudinal pitch requires each cover plate to be 2 feet 8 inches long. Fig. 49.

Let each line of $3\frac{1}{2}'' \times 3'' \times \frac{3}{8}''$ longitudinal angle bars be in two lengths 23' 4'' and 20' 8'' and cover the joint with a plate 2' 8'' $\times 3'' \times \frac{1}{2}''$ taking four outer table rivets on each side of the joint, the position of these covers being upon the underside of the top table, and upon the upperside of the bottom table.

Flush rivet at each end of the bottom table a 2′ 0″ × 1′ 3″ × ½″ sole plate *c*, Fig. 49, for distributing the weight of half the load over a sufficient surface upon each abutment.

It has been shown (**22**) that through diagonal action in the web the vertical stress of the load produces the horizontal stress which has to be resisted by the tables. The latter must in the present example, therefore, be imparted to each table by the rivets passing through the outer or horizontal tables of the 3″½ × 3″ × ⅜″ angle bars. Now it has already been seen that three $\frac{5}{16}$″ plates are necessary at the centre of each table, also that a cover plate for a joint in any one of these plates requires sixteen rivets on each side of the joint. Then for conveying the horizontal stress to the table plates the number of rivets required is 16 × 3 or 48, and adding 6 for the two angle bars, 54 rivets, to properly connect the web with each table, or 27 rivets in each line of angle bars, that is 9 feet run, whereas the effective run of riveting is 20 feet.

It is thus evident that the outer lines of rivets in each table do not serve for this purpose, and that these lines of ¾″ rivets to a 4″ pitch are adopted simply to bring the plates into efficient contact with each other throughout, so as to impart thereby rigidity to the whole structure. It is also evident that if each table were in one length and in one thickness of plate, the two outer lines of riveting might be dispensed with, excepting the two rivets required at the top and bottom end of each of the 6½″ × 3″ × $\frac{5}{16}$″ vertical T bar stiffeners hereafter described. The arrangement of table plates is therefore open to practical modifications in accordance with the exigencies of each case.

106. Web Details. Theoretically the horizontal area of the section of the web for the length extending from the centre of the span to an abutment would require to be the same as that of a table (**103**), and therefore in this instance 13.2 inches (**105**). Now as half the length of the span is 240 inches, the theoretical thickness of the web throughout would if the load were central $= \dfrac{13.2}{240} = .055$ inch.

But with a distributed load, as in the present example, the horizontal area of the half web would in order to resist vertical stress necessarily be the area of an acute-angled triangle extending from the centre of the span to either abutment (**26**), where its base in this example would $= .055 \times 2 = .11$ inch or the extreme thickness of the web.

Further .11 multiplied by the depth of the web in inches, and by the coefficient will give the shearing resistance of the web at an abutment.

$$\text{Thus } .11 \times 70 \times 6\tfrac{1}{2} \text{ tons} = 50 \text{ tons}$$

and 50 tons is half the weight of the distributed load.

Fig. 50 is a diagram of a plan of the web with the thicknesses drawn full size, and *c* being the centre of the span, the two acute-angled triangles *a c b* and *a' c b'* when measured by the scale of $\frac{1}{8}''$ to the foot for length, and full size for width, will give the theoretical horizontal area required for the web, the base *a b* and *a' b'* of each triangle being .11 inch.

But in the first place it would be impossible to readily get plates of tapering thicknesses, and even if such plates could be obtained it is evident that a thickness of .11 inch tapering to nothing at the centre of the span could not be practically adopted when wear and tear and contingencies are taken into

consideration. Besides, if it should be necessary or desirable
to form the web longitudinally of a series of plates, its thickness should then be sufficient to present an adequate bearing
area to the rivets for resisting the vertical stress of the load
at all vertical joints.

In order to comply with conditions laid down in Table 5 (**95**),
divide the web horizontally into four plates each 11 feet long,
Figs. 47 and 50, and let suitable covers be provided for the
joints. Then, as there can be no shearing strain at the central
joint from an equally distributed load, proceed to determine the necessary thickness of the two central plates and
the number of rivets necessary for a proper rivet-bearing
surface at each of the intermediate joints d, d. Now 50
tons being the load on the half span, and the length of
half the span being 20 feet, the vertical stress at each of
those joints $= \dfrac{50 \times 11}{20} = 27.5$ tons, and $\dfrac{27.5}{6.5} = 4.23$ inches,
or the bearing area required in the web plate for each vertical
line of rivets at the joint, and as the rivets are $\frac{3}{4}''$ diameter, the
length of bearing area required $= \dfrac{4.23 \times 4}{3} = 5.64$ inches..

Let these web plates be $\frac{1}{4}''$ thick then $\dfrac{5.64 \times 4}{1} = 22\frac{1}{2}$ rivets
required, or say 23 rivets.

By placing the centre of the end rivets 2" from the top and
bottom of the plates, the intermediate rivets will coincide with a
3-inch pitch. But in any case, a requisite fractional deviation,
however small, from any given regular 4 or 3 inch or other
pitch, is under no circumstances a matter of any difficulty in
the work. Moreover, many contractors are provided with
drilling and punching *Tables*, or benches* upon which a plate

---

* The invention of Mr. Field, Messrs. Maudslay, Sons, and Field.

may be brought under the drill or punch in a straight line of successive distances of any length of gauge or pitch.

In order to give a proportionate practical increase to the remaining portions of the horizontal area of the web let the plate at each end be $\frac{5''}{16}$ thick ; cover on each side of the web each of the three vertical joints with a $6\frac{1}{2}'' \times 3'' \times \frac{5''}{16}$ ⊤ bar, *e*, Figs. 47 and 48, also rivet to the web as stiffeners a pair of similar ⊤ bars *f* at each end of the span immediately over the abutment, and a pair of $3'' \times 3'' \times \frac{5''}{16}$ angle bars, *g*, Fig. 47, at each end of the web ; crank and rivet all these at top and bottom to form stiffeners for the tables as shown Fig. 48.

The *longitudinal* dimensions given throughout will all fall in with a 4-inch pitch of riveting.

---

107. Constructive Deviation from Theoretical Lines in some instances becomes necessary in order practically to meet special requirements.

For instance, suppose a *girder bridge* be required to carry at a given level an ordinary 40-feet wide roadway across a stream also 40 feet wide, so as to leave for the latter a stipulated clear headway, above which to the surface of the proposed road there would then remain a depth of 4 feet 3 inches available for the platform of the bridge.

Allow a depth of 1 foot 3 inches above the girders for filling and road metal, thus leaving 3 feet only or 1 to $13\frac{1}{3}$ of span for the depth of girders when placed longitudinally beneath in the direction of the road.

As a simple and durable mode of construction let the road-

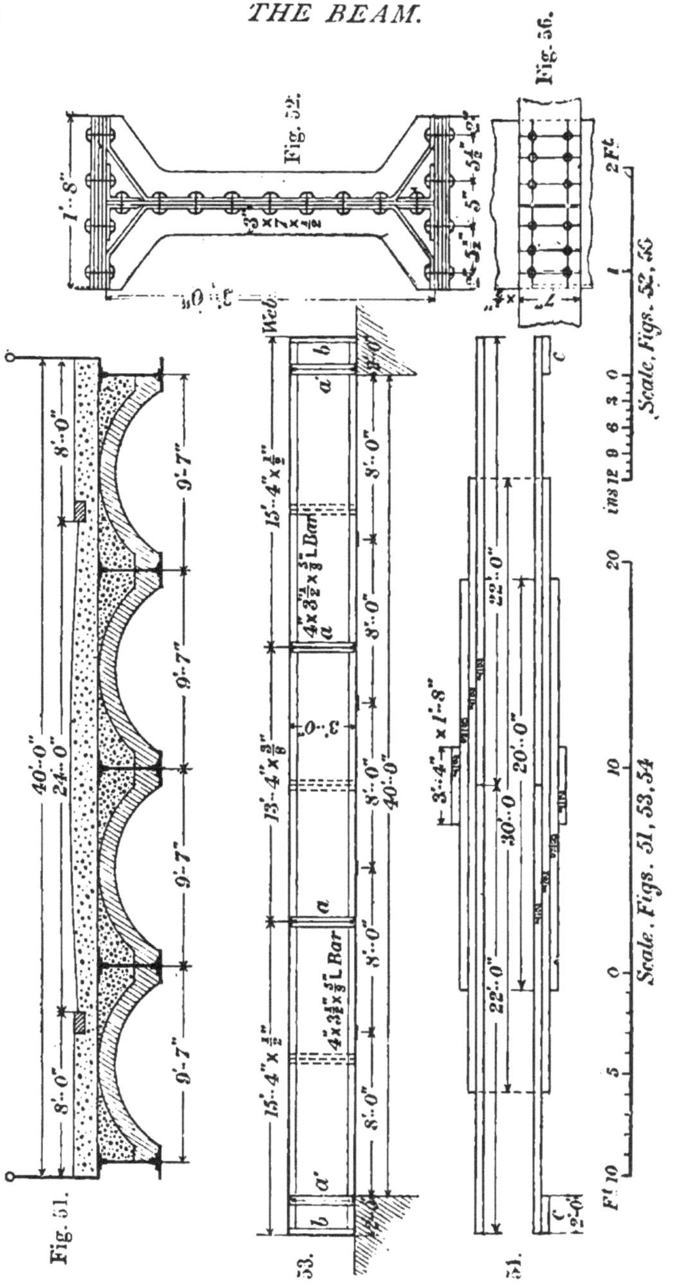

Fig. 52.

Fig. 56.

Fig. 51.

53.

54.

Scale, Figs. 52, 56.

Scale, Figs. 51, 53, 54.

way be supported on brick in cement jack arches, springing from the lower tables of the girders, with their spandrils filled in with Portland cement concrete, as shown in transverse section Fig. 51.

The next step will be to determine the distance apart of the girders from centre to centre. Now when arrangements permit, it is desirable that the versed sine of a segmental jack arch should not be less than one-fourth of its chord, for the flatter the arch the greater will be the horizontal crushing stress at the crown and the thrust at the springing. The maximum rise of arch, therefore, as fixed by the available depth may practically, to some extent, limit its span, and thus the distance apart of the girders. Thus with arches of two rings of brick or 9 inches thick, and by a division of the platform of the bridge into four 9 feet 7 inches bays as shown, Fig. 51, the above conditions will be well covered.

But the most efficient arrangement for the girders themselves is that by which a given uniformly distributed live load upon the fraction of the area of the platform carried by each girder shall cause the same stress within the girder as that which would be caused by a given maximum incidental rolling load which it may have to carry. For then the girder need not have an excess of strength in order to carry either of those loads which would be comparatively useless in the case of the other.

Take the maximum rolling load as 32 tons on four wheels 10 feet longitudinally and 4 feet transversely apart (**65**).

Then when this load incident in four equal fractions of 8 tons is passing the centre of the span the mean distance of each fore and aft wheel from that central line will be 5 feet, and the

half span being 20 feet the effect at the centre of the span
will be—

$$(4 \times 8 = 32) \times \frac{15}{20} = 24 \text{ tons.}$$

But as the width of the wheels apart is 4 feet their mean
distance from the centre of a girder may be taken as 2 feet,
and 9 feet 7 inches having been determined as the distance
apart of the girders, from centre to centre, the incidence of
the load upon an intermediate girder will be —

$$\frac{24 \times 7' \ 7''}{9' \ 7''} = 19 \text{ tons.}$$

Therefore 19 tons at the centre of the span will be the
effect of the assumed maximum rolling load upon the
girder.

For the required equivalent distributed or live load per
foot super (**29**), 19 tons at the centre of a girder is equivalent to or will cause the same horizontal stress as 38 tons or
760 cwt. longitudinally distributed over the span of 40 feet
and the width of 9' 7" carried by the girder,

$$\text{then} \ \frac{760}{9.58 \times 40} = 2 \text{ cwt. per foot super.}$$

If the platform were divided into three bays instead of four
with only four girders, the distributed load efficiency would
be reduced to $1\frac{1}{2}$ cwt., the allowance given, page 80.

But the proposed arrangement is practically preferable,
for, with three bays only, the proportionate depth of the
versed sine of each arch would be reduced, and this would
be further diminished by the necessity with such an increase
of span of three instead of two rings of brickwork for the
arches.

In order to determine the constructive details of these

girders, first ascertain the amount of a central load equivalent to the loading they are intended to carry. This, for an intermediate girder, will be,

The maximum rolling load as determined . . 19 tons
For brickwork, filling, and roadway
  20′ 0″ × 9′ 7″ × 3′ 0″ = 575 ft. cube @ 1 cwt. 29 ,,
For the weight of half a girder, say . . . 2 ,,
  ⎯⎯
Total at centre of girder, say . 50 ,,

108. Table Design. Again assuming steel as the material to be used, to find the efficient sectional area for each table at the centre of the span. Formula (21), (**76**),

$$a = \frac{w\, l}{4\, d\, c} = \frac{50 \times 40}{4 \times 3 \times 6\frac{1}{2}} = 25.64 \text{ inches area.}$$

Let two 4″ × 3½″ × ⅝″ angle bars be provided as a suitable connection between the web and each table, and let ⅞″ rivets with a 4-inch longitudinal pitch be used throughout, as these girders, excepting the lower face of the bottom table, are to be covered in permanently with brickwork and concrete.

Take the area of the two outer flanges of these two angle bars, less rivet holes, as a part of the total area required for each table.

Then (4″ − ⅞″ = 3⅛″) × 2 × ⅝″ = 3.9 inches area and 25.64 − 3.9 = 21.74 inches, or the central sectional area to be provided by the table plates.

Let the width of each table be 20 inches, from which deduct the width of four ⅞″ rivet holes. (Fig. 52.)
Then 20″ − (⅞″ × 4) = 16½″ wide, and $\dfrac{21.74}{16.5} = 1.32″$ thick, say 1⅜″ thick, and 16½ × 1⅜ = 22.69 area.

With this liberal allowance for the total thickness of plates adopt as a suitable arrangement of thicknesses one $\frac{3}{8}''$ and two $\frac{1}{2}''$ plates for the central part of each table.

Referring to (**37**), it will be seen that horizontal stress set up in the tables of a girder by a concentrated load moving over a span decreases from its centre to each support in the same ratio as that caused by an equally distributed stationary load. Further (**41**), that with an uniformly distributed rolling load, the stress to be resisted by the tables is a maximum at every point in the span when the load has advanced sufficiently to cover the whole.

The lengths of the table plates, as shown in diagram, Fig. 54, have therefore been arranged to practically meet the proportionate stress at any part of the span as determined for an uniformly distributed load (**27**).

Let the inner line of plate of each table be in two 22-feet lengths connected with a cover plate.

Now as the transverse run of section of each plate less rivet holes $= 16\frac{1}{2}$ inches, that should also be the aggregate sectional run of rivets passing through the cover plate on each side of the joint (**101**), or just upon 19 rivets are required. But as the rivets are arranged transversely in rows of four, take five rows or 20 rivets which with a 4" longitudinal pitch require a cover plate 3' 4" long, and $\frac{1}{2}''$ thick, Fig. 54.

Let each 4" × $3\frac{1}{2}''$ × $\frac{5}{8}''$ angle bar be in two convenient lengths and cover the joint with a 2' 8" × 4" × $\frac{5}{8}''$ plate taking on each side of the joint four of the outer rivets of the table in order to carry on the efficiency of the outer flange of the angle bar.

Flush rivet at each end of the bottom table a 2' 0" × 1' 8" × $\frac{1}{2}''$

solo plate *c*, Fig. 54, for transferring half the weight of the load to a bed stone built into each abutment.

Now the area of each sole plate = $3\frac{1}{3}$ feet, which multiplied by 20, the coefficient given for York stone, Table 2 (**63**) = $66\frac{2}{3}$ tons, thus leaving a good margin of strength, as the total stress at an abutment is 50 tons (**107**).

109. Web Design. As in the present example the required area of each table is 25.64 inches, that will also theoretically be the required horizontal sectional area of the half web extending from the centre of the span to an abutment (**103**). Thus as this length is 20 feet or 240 inches, the theoretical thickness of the web throughout would if the load were central = $\dfrac{25.64}{240}$ = .107 inch.

Therefore, as before shown (**106**), with the same load equally distributed the horizontal section of the half web would be the area of an acute-angled triangle extending from the centre of the span to either abutment, where its base in this example would = .107 × 2 = .214 inch, and .214 multiplied by the depth of web in inches and by the co-efficient will give the shearing resistance in the web at an abutment.

Thus .214″ × 36″ × $6\frac{1}{2}$ tons = 50 tons,
and 50 tons is half the weight of the total distributed load (**26**).

Fig. 55 (next page) is a diagram plan of the web with the thicknesses drawn full size, and *c* being the centre of the span the hatched triangle *a c b* represents the theoretic section of the half web required for the uniformly distributed load of 50 tons carried by the half girder, while the figure *a d e b* represents the theoretic section required should the same load move or roll over the span. Series A and B represent proportionately the vertical stress set up in the web by first the *distri-*

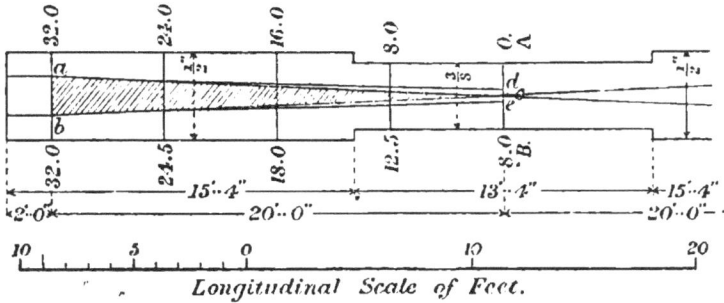

Fig. 55.

buted and second the *rolling* condition of the load (**40**). Thus under the first condition the thickness of the web at the abutment having been found to be .214 inch it will be nil at the centre of the span, while under the latter condition the thickness at the abutment will remain the same; but that at the centre of the span will be one-fourth of that at the abutment or $\dfrac{.214}{4} = .0535$ inch.

Therefore with a load of the same weight, either central, or uniformly distributed over the span, or in the form of a concentrated rolling, or of a distributed rolling load, in each case with the same depth of beam, the theoretic horizontal area and consequent weight of the web would be in the following ratio:—

When the load is
$$\begin{cases} \text{Central} & . & . & . & 12 & \textbf{(23)} \\ \text{Distributed} & . & . & . & 6 & \textbf{(26)} \\ \text{Concentrated rolling} & . & 18 & \textbf{(36)} \\ \text{Distributed rolling} & . & 7 & \textbf{(40)} \end{cases}$$

Now in these figures, the whole of the load is assumed to be either central, distributed, or rolling. But that is not so in the present example, for out of a load of 50 tons, 19 tons only form a concentrated rolling load.

Thus theoretically the thickness of the web will at an abutment remain the same as before, and the thickness at the centre of the span will $= \dfrac{.0535 \times 19}{50} = .02$ inch.

But it is evident that the adoption of such thicknesses in practice would as already stated (**106**) be simply inadmissible, especially as these girders are to be embedded in brickwork and concrete, and thus cannot subsequently be repainted.

Therefore make the central part of the web say $\frac{3}{8}''$ thick, and the remainder $\frac{1}{2}''$ thick, as shown by the outer lines in the plan of thicknesses. Fig. 55.

The outer or face girders having a less width of platform to carry should have a proportionately reduced table area, but in practice the web may be the same for all the girders.

Cover the joints of the web plates on each side of the web with a $7'' \times 3\frac{1}{2}'' \times \frac{1}{2}''$ vertical **T** bar at a, a, Fig. 53, and provide similar bars at $\dot{a}, \dot{a}$, to assist in transferring the vertical stress of the load to the abutments, also a pair of $3\frac{1}{2}'' \times 3\frac{1}{2}'' \times \frac{1}{2}''$ vertical angle bars, b, b, at each end of the web. All these should be cranked as shown, Fig. 52, and each bar connected at top and bottom with the tables by two of the outer line of table rivets.

Construct the web of each face girder with similar and also intermediate **T** bars, shown by dotted lines in the figure, for the better security of iron, stone, or brick parapets.

For the vertical riveting of the web, as $\dfrac{13' \ 4''}{2} = 6' \ 8''$, Fig. 53, the vertical stress at either of the joints at a, a, $= \dfrac{50 \times 6.66}{20} = 16.66$ tons, and $\dfrac{16.66}{6\frac{1}{2}} = 2.56$ inches area required.

With $\frac{7}{8}''$ rivets a *length* of $\dfrac{2.56 \times 8}{7} = 2.93$ inches bearing surface is required.

Thus with a $\frac{3}{8}''$ plate $\dfrac{2.93 \times 8}{3} = 7.81$ rivets are required.

And nine rivets on each side of the joint as given by a $1''$ pitch are quite sufficient.   Fig. 52.

110.   Tie-bars should be provided, say at distances 8 ft. apart, or $\frac{1}{5}$th of the span, to relieve the girders of the lateral thrust of the jack arches. Now as the distributed dead load on half the span of an intermediate girder $= 50 - 19 = 31$ tons **(107)**, and is statically equivalent to a similar central load on the crown of an arch, that taken by each tie-bar would be $\dfrac{31}{5} = 6.2$ tons, add to this 8 tons for the maximum live load which might be caused by one wheel of the loaded truck **(107)** passing longitudinally along in the line of the crown of an arch :

Then $6.2 + 8 = 14.2$ tons, or the vertical stress on the crown of each arch to be resisted by each tie-bar.

Now 14.2 tons on the centre of an arch $= 7.1$ at each abutment, and as the versed sine of the arch is to half the chord as 1 to 2, the tensile strain on a tie-bar is $\dfrac{7.1 \times 2}{1} = 14.2$ tons, and therefore as the coefficient for steel is 6.5 tons per square inch $\dfrac{14.2}{6.5} = 2.18$ inches area required in each tie-bar.

Let these tie-bars be $7''$ wide $\times \frac{1}{2}''$ thick, and attached at each end to the under side of the bottom table of the girders by four of the table rivets plus two $\frac{7}{8}''$ intermediate rivets, Fig. 56. It will then be found that these bars and the six $\frac{7}{8}''$ diameter rivets give amply sufficient tensile, compressive, and shearing area for the resistance of those stresses, also that the efficient sectional area taken from the table by.

the additional rivet holes is fully covered by the tie-bar itself.

There will be the same amount of horizontal stress, or 14.2 tons, acting in compression at the crown of the arch. Assume this to be distributed by the arch over 4 feet on each side of a tie-rod, viz. over a length of 8 feet, or the distance apart of the tie-bars.

Now the sectional area of that length of arch is 8′ 0″ × 9″ = 6 feet, and $\frac{14.2}{6}$ = 2.37 tons compression per square foot, whereas by Table 2, (**63**), 8 tons is a safe compressive load for brick in Portland cement. It should, however, be remembered that a rolling load when in motion may cause somewhat serious abnormal shocks and consequent strain.

111. Comparative Weight of the two girders described (**104, 107**), each being of the same span and designed to carry the same load.

Let the first, or girder with the deep web, be A and the other B.

Taking the horizontal flange of the longitudinal angle bars as a part of each table, and the vertical flange of the same bars as a part of the web, also adding for rivet heads two per cent. of the total weight of plates and bars, the following are the results when taken out and computed from the given dimensions:—

	A.	B.
Web . .	1.92	1.82
Tables . .	1.84	3.60
Total .	3.76	5.42 tons.
or	7	to 10.

It will thus be seen that when compared with B there is

30 per cent. less material in **A**, and that the weight of the **A** web is only $\frac{1}{10}$ ton, or **2** cwt. more than the **B** web.

The weight of the deep web happens to be partly kept down by the longer run of the thinner plates; see Figs. 50 and 55.

The weight of the tables will be found to be inversely as the depth of each girder.

112. Wind Trussing for the resistance of strains set up by gales should invariably be a matter of consideration in any designs for exposed structures, especially in the case of those which would present large surfaces to wind pressure.

For instance, the distributed lateral wind pressure upon the girder described (**104**), when computed by the Board of Trade factor of half a cwt. per foot super, would be 6 tons.

How the necessary resistance is or is to be provided for must entirely depend upon the nature of each case. In that of the bridge described (**107**), the structure requisite for carrying the dead and live loads provides at the same time, as will be evident, ample resistance also to any possible lateral wind pressure.

In some cases the requisite resistance may be provided in connection with the diagonal platform trussing of a bridge, or with that supplemented by overhead trussing.

CHAPTER X

113. The Application of the Principles of the Parallelogram of Forces (**15**) is the readiest method of determining the stresses set up in the members of these girders by the direct stress of any kind of load.

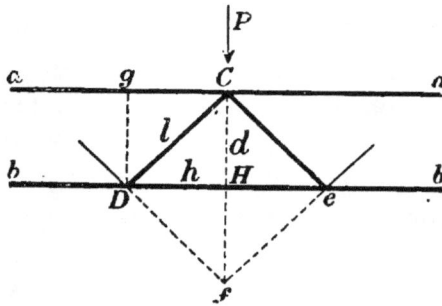

Thus let the parallel lines $a\,a$ and $b\,b$, Fig. 57, represent portions of the top and bottom tables of a framed girder, and $C\,D$, $C\,e$, the two equal diagonal members of a bay $D\,e$ of the web. Assume that a load or pressure

Fig. 57.

P be applied at the point C acting vertically downwards in the direction of the vertical line $C f$, draw $D f$ parallel to $C e$, and $e f$ parallel to $C D$. Then $C\,D f\,e$ is a parallelogram, and if the length of the line $C f$ be taken to represent the amount of the pressure P, so will the length of the line $C D$

or C e also represent the compressive stress set up by the pressure P in each of those members. Draw the line g D parallel to Cf. Then as g D $=$ C H, and g C $=$ D H, g C D H is a parallelogram of which C D is a diagonal.

Now as the length of the line C D has been found to represent the compressive stress upon that strut, so will the length of the line g D represent the vertical stress at the point D, caused by the strut C D, and as the line D e the horizontal diagonal of the parallelogram C D $f e$ is bisected in the point H by the vertical diagonal Cf, so the length of the line D H will represent the amount of tensile stress set up between the points D and H by the pressure P in the bottom table $b\,b$ of the girder, and H e will in the same way represent the necessarily equal and opposite amount of the same stress. Then g C being equal to D H represents an equal amount of thrust in the top table at the point C, which also is placed in a state of stable equilibrium by an equal and opposite thrust at the same point.

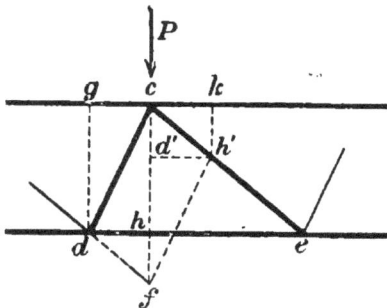

Fig. 58.

The mode of proceeding will be precisely the same if the diagonal web bars are of unequal length and meet the line of pressure of the load at unequal angles.

Let $c\,d$ and $c\,e$, Fig. 58, be two diagonal web bars of unequal length meeting at c in the top table of a girder, and let a pressure P be acting downwards at that point in the direction of the vertical cf. Draw df parallel to $c\,e$, and $f\,h'$ parallel to $c\,d$.

Now $cdfh'$ is a parallelogram, and if the length of the line cf be taken to represent the pressure P, then cd will be the compressive stress in that bar, and ch' that in the bar ce.

Complete the two subsidiary parallelograms $gcdh$ and $ckd'h'$. Then $d'h' = dh$, and as dh represents the tensile stress, so $d'h'$ represents the equal corresponding tensile resistance caused by the pressure P in the bottom table within the length of the bay de.

So also does gc represent the thrust of action, and kc that of equal reaction at the point c in the top table of the assumed girder.

114. The most Efficient Angle for Web Bars. Referring to Fig. 57, let $d = $ C H the depth of the girder, $l = $ C D the length of a bar, and $h = $ D H the horizontal length of a bar.

Now as any given bar increases in length so will it increase in weight, but further the parallelogram of forces has shown that with reference to the members of a framed web the stress upon any bar increases in a direct ratio with any proportionate increase of length.

Thus for any increase of length there should be throughout the bar a corresponding increase of strength, and therefore the weight of the bar will vary as the square of the length, or l^2.

Now the lines h and d in the figure may be taken to represent the two arms of a bent lever D H, H C, or two elements in the determination of stresses in a girder (**24**), and by the principle of the lever founded on the equality of moments, H being the fulcrum and the system being in equilibrium, either h or d may be taken as a pressure applied at the end of the other arm tending to turn the system about H. Therefore $h\,d$ = the moment of the system. And as h and d contain the right angle of the triangle C H D of which l is the hypotenuse $h^2 + d^2 = l^2$. Therefore if $h = 1$ the result must be a minimum when $d = 1$ also, and $h + d = 2$. Then $l = \sqrt{2} = 1.414$, and with a given amount of material, the most efficient angle for the bars of a web is 45° made by each side of the triangle C H D with the hypotenuse l, because then $\dfrac{l^2}{h\,d}$ is the minimum.

115. Useful Length Ratio for Web Bars. It has been seen that stresses in the web of any girder when resolved are relatively as the sides and hypotenuse of a triangle, also that when two equal sides contain a right angle and each thus forms an angle of 45° with the hypotenuse the proportion between the hypotenuse and either side is as $\sqrt{2} = 1.414$ to 1, or a fractional ratio.

But if we take a right angle contained by two lines relatively = 3 and 4, then the hypotenuse will = $\sqrt{(3^2 + 4^2)} = 5$, or a whole number for the length of a diagonal bar.*

These proportions of 3, 4, and 5 by eliminating fractions are therefore more simple in application, and at the same time definite.

---

* A workman's mode of accurately setting out from a point terminating a straight line, which cannot conveniently be extended beyond that point, a line at right angles with the given line, is to describe from the point with a radius of 4 units an arc on either side of the line, then from the given point to set off on the given line at a distance of 3 units a second point, and from it with a radius of 5 units to describe an arc cutting the first arc. Then a line drawn from the point of intersection of the two arcs to the given point will be at right angles to the given line. Vide " *the carpenter's,*" Proposition xlvii., Euclid, Book 1.

As the weight of the tables will remain constant, let us compare per unit of length the theoretical weight of a girder A constructed with the diagonal bars at an angle of 45° when h and d each equal 4, with that of B in which h, d, $l=$ relatively 3, 4, and 5.

Let w be the proportionate weight of the girder, then

$$w = \frac{l^2}{h\,d}$$

$$\text{Girder A,} \quad w = \frac{4 \times 4 \times 2}{4 \times 4} = \frac{32}{16} = 2.00$$

$$\text{,, B,} \quad w = \frac{5 \times 5}{3 \times 4} = \frac{25}{12} = 2.0833$$

$$\text{Thus if B} = 100 \text{ then A} = \frac{2 \times 100}{2.0833} = 96$$

or there would be 4 per cent. less material required in the first case than in the second.

But the second proportions possess for practical use the advantage of being not only definite, but at the same time they have in work a more pleasing appearance than the right angles given by the former.

As an instance, *a roof* the sides of which form a right angle at the ridge is æsthetically hard-featured. For slates therefore a roof might with advantage be constructed with that angle somewhat greater, while for tiles it should be less than 90°.

In the following nine examples the same conditions of span, depth, and loading have been adopted in order the better to illustrate variations of stress, relatively set up by various arrangements of web members. The thick lines in the diagrams are struts and the thin lines tie-bars.

Load Central.

116. Given a Single System Square and Diagonal Framed Girder 30 feet long between the supports, 4 feet deep, and carrying a central load $P = 80$ tons, Fig. 59. For the web, let the ratio of the lengths h d and l be 3, 4, and 5 (**115**).

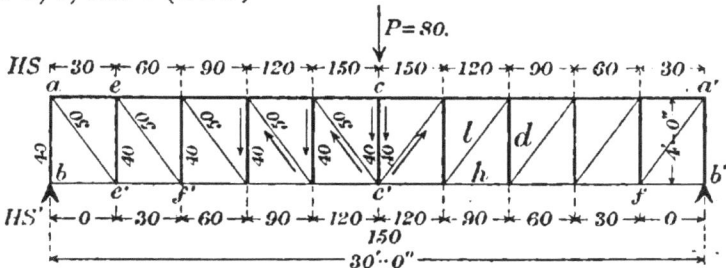

Fig. 59.

In this example the vertical lines of the diagram are struts, each $= d$ dividing the length of the girder into ten bays, each $= h$ in width, and the diagonal lines are tie-bars each $= l$, these forming in pairs the single system web.

Now as each bay is a parallelogram, and as the direct vertical stress caused by the load $P = 80$ is halved in the central vertical c c' (**23**), and each half $= 40$ is transmitted in an opposite direction through the verticals and diagonals of the web in the directions of the arrows to each support, there must necessarily be a vertical compressive stress $= 40$ in each of

the former as figured in the diagram, excepting the central
strut $c\ c'$ which has to resist the whole stress of the load.

The tensile stress in each diagonal tie-bar will therefore by
the parallelogram of forces

$$= \left(\frac{P}{2} = 40\right) \times \frac{l}{d} = 40 \times \frac{5}{4} = 50.$$

It will be seen from the diagram that a stress of 50 in
each tie-bar will place the top table $a\ a'$ in a state of horizontal compression, and that a stress of 40 in each strut
will place the bottom table $c'f$ in a state of horizontal tension.
Parts $b\,e'$ and $f\,b'$ of the latter might, however, be added,
simply as stays or steadiments, or the girder might be supported at a and a'.

To determine the *horizontal* stresses in each table: as $a\,e\,c'f'$
is a parallelogram, the stress in the first bay $a\ e$ of the top
table and in the first bay $e'f'$ of the bottom table $= 40\,\dfrac{h}{d} = 30$
or $50\,\dfrac{h}{l} = 30$. Proceeding in the same way to the central
vertical there will be an additional increment of 30 in each
bay plus 30 at the central vertical in the case of the bottom
table due to the two tie-bars which meet at c'.

The latter increment will, however, be taken by the pin or
rivet connecting these bars, and not by the table itself, unless
each bar should have a separate attachment.

The stresses as determined for each table are given Fig. 59,
Series HS, IIS'. These, as in the case of any ordinary
beam of uniform depth carrying a central load, increase from
each end to the centre of the span in terms of arithmetic progression (**24**).

M

Load Distributed.

117. Given a Single System Square and Diagonal Framed Girder 30 feet long between the supports, 4 feet deep, with vertical and diagonal web members, and carrying at the bottom table a distributed load of 160 tons or $5\frac{1}{3}$ tons per foot run, Fig. 60. For the web, let the ratio of the lengths h, d, and l be 3, 4, and 5 (**115**).

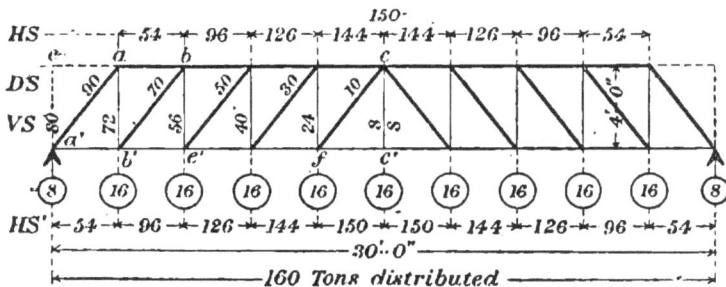

Fig. 60.

In this example the vertical lines of the diagram are tie-bars dividing the length of the girder into ten bays as in the last, and the diagonal lines are struts.

Now as the span is divided into ten bays, the stress of $\frac{1}{10}$ of the load = 16 will become active at each vertical, and the weight of one-half of the distributed load carried by each outer bay = $\frac{16}{2}$ = 8 will pass horizontally direct through the bottom table to the adjoining abutment. It is further evident that one-half of the fraction of the vertical stress caused by the distributed load = $\frac{16}{2}$ = 8 carried by the tie-bar at the central vertical will be transmitted by the members of the web to each support + 16 at each intermediate vertical, and + 8 at the support itself, at which the stress will thus equal 80 or half the amount of the total weight of the load, Fig. 60, Series VS.

Thus, commencing from the central vertical, let s be the stress in a tie-bar, and t the thrust in the consequent strut, then $s\dfrac{l}{d} = t$, or for the strut $c\,f$, $t = 8 \times \dfrac{5}{4} = 10$, and so on for each successive strut to the abutment plus the thrust as determined for the last. Fig. 60, Series DS.

For the *horizontal* stresses in each table, as $a\,b'$ is the diagonal of the parallelogram $a\,b\,a'\,b'$, the stress from a to b and from a' to $b' = s\dfrac{h}{d}$, or in this example $= 72 \times \dfrac{3}{4} = 54$ and so on to the central vertical, but in each case adding the product of the last, thus for the next $(56 \times \dfrac{3}{4} = 42) + 54 = 96$. For the central vertical, it will be remembered, that in this example s will equal 8.

Series HS and HS′ give respectively the horizontal stresses in each table, and these increase towards the centre of the span as the multiple of the segments into which it is divided by the verticals, as in the case of an ordinary beam of uniform depth, carrying an equally distributed load (**27**).

The additional central increment of horizontal thrust = 6 at c, will be taken by the pin or rivet connecting the two struts which meet at that point, and not by the top table itself, unless each strut should have a separate attachment.

The continuation of this girder to square ends, as shown by dotted lines in the figure, would obviously add nothing to its efficiency, but might be made useful as a steadiment in connection with a wall or other lateral support.

Load Central.

118. Given a Single System Diagonal Framed Girder 30 feet long between the supports, 4 feet deep, and carrying a central load $P = 80$ tons, Fig. 61. For the web let the ratio of the lengths h, d, and l be 3, 4, and 5 (**115**).

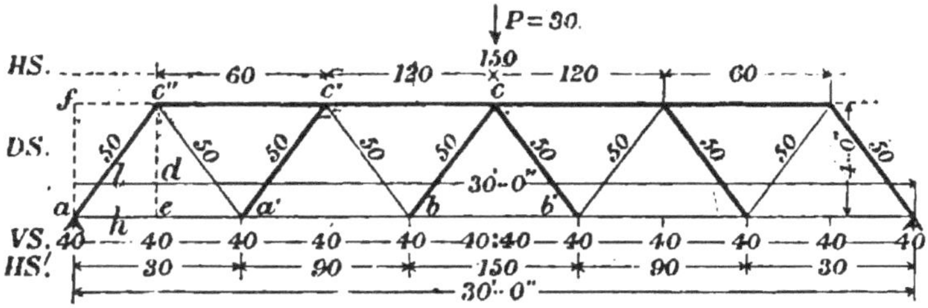

Fig. 61.

Thus the girder is longitudinally divided by the members of the web into 5 bays, each $= 2\,h$, or in this example 6 feet long, the six diagonal struts and the four diagonal tie-bars together forming the single framed web.

Now *vertical* stress caused by the load P is halved at the central point c (**23**), and each half $= 40$ tons is transmitted in an opposite direction through the diagonals of the web to each support, becoming active at each upper and lower junction of the web diagonals and tables. Series VS.

The strain in each diagonal will therefore be

$$40\,\frac{l}{d} = 40 \times \frac{5}{4} = 50 \text{ tons,}$$

as figured in the diagram Series DS.

To determine the *horizontal* stresses in the tables, assume in the first place the line $c'\,c''$ to be extended to f, and let

$f\,c'' = 3 = a\,e.$ Then because the resistance of the support a is vertical in the direction $a\,f$ and as $f\,c''\,a\,e$ is a parallelogram of which the diagonal $c''\,a = l$ and $a\,e = h$, and as there is a compressive stress of 50 in the direction of $c''\,a$, there will be a tensile stress in the bottom table from a to e, and consequently from a to the central vertical equal to $50\,\dfrac{h}{l} = 30.$

Next as $c''\,c'\,a'\,b$ is a parallelogram, the diagonal of which $c'\,a' = l$ and $c''\,c'$ or $a'\,b = 2\,h$, the compressive stress from c'' towards the central vertical $= 50\,\dfrac{2\,h}{l} = 60$, and likewise the tensile stress from a' to the centre will also be 60, but $+\,30$ already obtained for the bay $a\,a'$, and $60 + 30 = 90$ the total stress in $a'\,b$. For the central bay $b\,b'$ of the bottom table the tensile stress will in the same way be 60, and $60 + 90 = 150$ tons. It has been shown that the compressive stress due to the outer bay $c''\,c'$ of the top table is 60, therefore for each of the like two central bays it will be $60 + 60 = 120$ tons or 30 tons less than the stress in the central bay of the bottom table. The two struts, however, meeting at c cause a stress at that point $= 50\,\dfrac{h}{l} = 30$, and $120 + 30 = 150.$ But the last increment would be resisted by the pin or bolt connecting the two struts at c, and not in any way by the table itself, unless those two members should each be separately connected with it. Fig. 61, Series HS and HS′.

As a horizontal stress of 60 tons is imparted to each table at each junction of the tensile diagonals, the pin or bolt forming that junction should be so proportioned as efficiently to resist that stress, being the maximum to which it would be subjected in any one direction, for the diagonal stress has been found to be 50 and the vertical 40 tons.

Load Central.

119. Given a Double System Diagonal Framed Girder 30 feet long between the supports, 4 feet deep, and carrying a central load P = 80 tons, Fig. 62. For the web let the ratio of the lengths h, d, and l be 3, 4, and 5 (**115**).

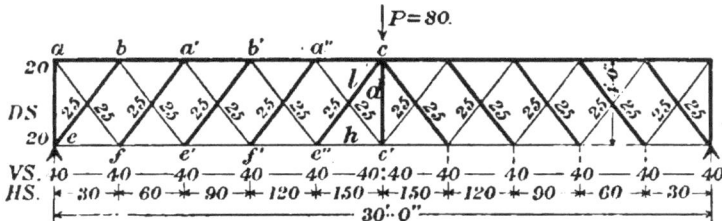

Fig. 62.

It will be seen from the diagram reading from the centre of the span that in this example there are two quite distinct and separate systems of diagonal struts and tie-bars, and tie-bars and struts, and that without the action of the central vertical strut $c\,c'$, the second system of diagonals would remain practically useless for transmitting weight from the load to the abutments.

But by the addition of that strut the efficiency of each of the two systems for the transmission of stress is rendered precisely the same, so that half the stress caused by the load = 40 will be resisted by the two diagonal struts meeting at c, while the strut $c\,c'$ will convey the remaining half = 40 direct to the point c', there to be taken by two diagonal tie-bars.

Now with the single system the girder was longitudinally divided into five 6-feet bays, giving a stress = 50 in each diagonal member, but as in this example the two systems divide the length of the span into ten 3-feet bays, with two active diagonals crossing each other in each bay, it will be evident that the stress in each diagonal will now $= \dfrac{50}{2} = 25$ tons. Series DS.

The compressive stress in each vertical end strut as $a\,e$ $= \dfrac{25 \times 4}{5} = 20$ caused by a diagonal tie as $a\,f$, while at the same time the strut $b\,e$ also brings a vertical weight at e upon the same support $= \dfrac{25 \times 4}{5} = 20$, and $20 + 20 = 40$ tons, or the weight of half the load, upon each abutment. Series VS.

To determine the *horizontal* stress to be resisted by the top table; as $a'\,f = 5$ is the diagonal of the parallelogram $a\,a'ff'$ and $a\,a' = 2\,h = 6$, the stress in $a\,a' = 25 \times \dfrac{6}{5} = 30$, in the same way as $b'\,c'$ is the diagonal of the parallelogram $b\,b'e'e''$, the stress in $b\,b' = 25 \times \dfrac{6}{5} = 30$, but as the parallelogram with the diagonal $a'f$ has also been found to give a stress of 30 between b and a', and consequently onwards to the centre c, the total stress in the bay $b\,a'$ of the top table $= 30 + 30 = 60$. So with each succeeding bay there will be an additional increment of 30, bringing the stress in the two central bays up to 150 tons, or the central stress already shown (**118**) to be set up in the same girder carrying the same load, but constructed with a single system web.

The *horizontal* stress resisted by the bottom table may be determined in precisely the same way, for as $b\,c'$ is the diagonal of the parallelogram $b\,b'e\,e'$, the stress in $e\,e'$ $= 25 \times \dfrac{6}{5} = 30$, and proceeding as with the top table there will be an addition of 30 in the bay $f e'$, and in every succeeding bay to the central vertical.

Horizontal stress in each table may also be determined simply by leverage. For as there is a vertical stress of 40 at e, the same stress will exist in each assumed vertical line coincident with the upper and lower angles formed by the web members, as $b f$ and $a'e'$, or 20 at b + 20 at f = 40.

Thus assume $a\,e\,f$ to be a bent lever with a fulcrum at e, and a stress of 40 in a vertical $b f$, then the stress in the direction $a\,b$ or $f\,e = \dfrac{40 \times 3}{4} = 30$, and so on with each bay, but + 30 at each succeeding vertical, bringing the stress in each of the two central bays up to 150 tons. Series HS.

It is usual to bolt or rivet the struts and tie-bars to each other at each crossing in order that they may impart *lateral* rigidity each to each, but this in no way affects the main conditions of the stresses.

If the central load were carried upon or suspended from the bottom table, the central strut $c\,c'$ would then become a tie-bar, but stress in all the other members would remain the same as before.

It will be observed that with a double system web and a central vertical member, vertical stress due to a central load becomes resolved into two equal diagonal cross stresses, the one of compression and the other of tension, as in the web of a plate girder (**22**). In this example, a diagonal compressive stress of 25 becomes further resolved into horizontal tensile stress of 30 in the bottom table, and likewise a diagonal tensile stress of 25 into horizontal compressive stress of 30 in the top table. The pins at the junction of the diagonals with the tables should therefore in this as in the last example and under all similar conditions have an efficiency equal to the major stress.

Load Central.

120. Given a Quadruple System Lattice Girder 30 feet long between the supports, 4 feet deep, and carrying a central load $P = 80$ tons, Fig. 63. For the web let the ratio of the lengths h d and l be 3, 4, and 5 (**115**).

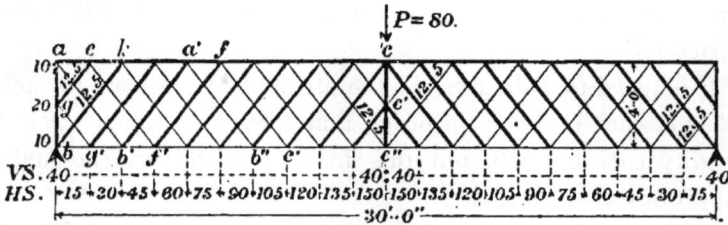

Fig. 63.

With this arrangement the girder is now divided into twenty bays, and the central strut c c'' by being connected with the central cross bars, and performing the same office as in the last example, will equally distribute vertical stress caused by the load or 80 tons between all the diagonal members of the web. Thus 20 tons will be taken by two diagonal struts at c, 20 by two half-length tie-bars at c', 20 more at the same point by two half-length struts, and the remaining 20 by two tie-bars at c'', and the vertical stress resisted by each diagonal strut and tie-bar is $\frac{20}{2}$ or 10 tons, therefore the direct longitudinal stress in each

diagonal member of the web $= \dfrac{10 \, l}{d} = 12.5$. Having thus
passed through the web members, each half of the direct
stress caused by the load is received by the vertical strut at each
end of the girder, and so transmitted to the supports, each of
these struts receiving 10 tons delivered by a tie-bar at its top,
20 delivered by a strut and tie-bar at its centre, and 10 delivered by a strut at its foot, or in all $10 + 20 + 10 = 40$ tons.

In commencing the determination of the *horizontal* stresses
to be resisted by each table it will be at once evident that
the two half members of the web, as $e \, g$ and $g \, g'$, at each of its
ends merely transfer diagonally vertical stress to the vertical
end strut, because in neutralising each other's *diagonal* energy
at the point of their incidence they convey no horizontal stress
whatever to either table.

The amounts of *horizontal* stress may be determined in
precisely the same way as in the last example. But as the
stress in each diagonal is now 12.5 or one-half of that given,
Fig. 62, so will be the primary result at each step of the
solution. The number of steps or bays is now, however,
double, and thus in reading from either end to the centre of
the girder the horizontal stress at the second, fourth, sixth,
&c., verticals in Fig. 63 is the same as that at the corresponding first, second, third, &c., verticals in Fig. 62.

Therefore for determining the horizontal stress in the top
table; as $a' \, b' = 5$ is a diagonal of the parallelogram $a \, a' \, b' \, b''$
and $a \, a' = 2 \, h$, the stress in $a \, a' = 12.5 \, \dfrac{2 \, h}{l} = 15$. In
the same way as $f \, f'$ is a diagonal of the parallelogram $e \, f \, f' e'$,
the stress in $e \, f$ also $= 12.5 \, \dfrac{2 \, h}{l} = 15$, but as the first
parallelogram has been found to give a stress of 15 from
a to a', the stress in the bay $e \, k$ must consequently be

$15 + 15 = 30$. So with each succeeding bay there will be an additional increment of 15, bringing the stress in the two central bays up to 150 tons.

Horizontal stresses in the bottom table are it is evident determined in the same way, and coincide in each bay with those in the top table. Series HS.

By comparing this with the two previous examples, it will be seen that the stress in the diagonal members of the web is inversely as their number. Also that in each case the horizontal stress in each table at the centre of the span is 150 tons, or the same that would exist at the central vertical of an ordinary girder with the same span depth and load.

For by the usual formula (21) (**76**), omitting coefficient c, let S be the stress at the centre of the span.

$$\text{Then } S = \frac{w\,l}{4\,d} = \frac{80 \times 30}{16} = 150 \text{ tons.}$$

Further it will be seen that the horizontal stress increases in terms of arithmetic progression from each support to a maximum at the centre of the span as in the case of any ordinary beam or girder carrying a central load (**24**).

Each theoretical half-length central diagonal strut and tie meeting at c' would practically be constructed as one whole length strut.

Load Distributed.

121. Given a Single System Diagonal Framed Girder 30 feet long between the supports, 4 feet deep, and carrying at the bottom table a distributed load of

160 tons or $5\frac{1}{3}$ tons per foot run, Fig. 64. For the web let the ratio of the lengths h, d, and l be 3, 4, and 5 (**115**).

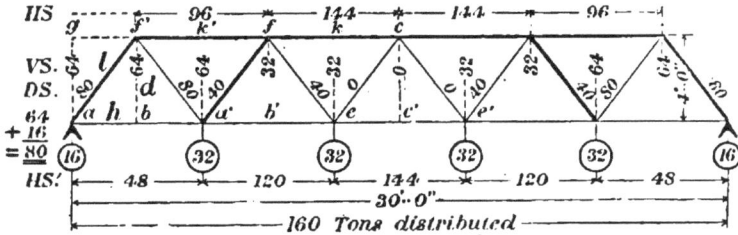

Fig. 64.

The members of this girder are in number, length, and arrangement precisely the same as those of that shown, Fig. 61, page 164. Thus the span being 5 bays in length, the proportion of the distributed load carried by each bay is $\dfrac{160}{5} = 32$ tons.

Now as the *vertical* stress caused by an evenly distributed load becomes equally divided at the centre of the span and transmitted through the medium of the web in opposite directions to the two points of support, the first step towards determining the stresses set up in the various members will be to apportion the amount of this stress to each point of the framing at which it becomes active.

First $\frac{1}{5}$ of the load $= 32$ extending from b' to c' becomes active at the point e, causing a tensile stress of $32 \dfrac{l}{d} = 40$ in the tie-bar fe, and consequently a compressive stress of 40 in the strut fa'.

No stress whatever will be caused by the load in the diagonal bar ce, or in the corresponding member ce' on the other side of the central vertical $c\ c'$, because in these, it is evident on inspecting the diagram, any tendency to tension or compression is balanced. Theoretically, therefore, these diagonals might be omitted altogether in the case of an evenly distributed load, except for the purpose of bracing.

Second $\frac{1}{5}$ of the load or 32 extending from b to b' becomes active at the point a', and as it has been shown that there is a stress of 40 in the strut fa', so there will be an additional vertical stress of $40\ \frac{d}{l} = 32$ at a'. The total vertical stress at that point is therefore $32 + 32 = 64$, and the total tensile stress in the tie-bar $f'\ a'$ and the total compressive stress in the strut $f'a$ is $64\ \frac{l}{d} = 80$. Draw the vertical $a\ g = d$ and $g\ f' = a\ b = h$, then as $g\ f'\ a\ b$ is a parallelogram of which $f'a$ is the diagonal, the weight brought to the support a by the strut $f'a$ is $80\ \frac{d}{l} = 64$, to this add $\frac{32}{2} = 16$, or the weight of the portion of the load extending from b to a and carried direct to the latter point by the bottom table, then the total weight upon the support a will be $64 + 16 = 80$, or half that of the load, the other half being taken in a similar way by the support at the other end of the girder, Series VS. The stresses in the diagonals are given in Series DS.

It has therefore been seen that the total vertical stress transmitted *through the web diagonals* to each support is 64 only, and not 80 tons, or the weight of half the load. The former stress is therefore alone active in setting up horizontal stress in the tables.

Horizontal compressive stress in the top table may be determined in the following way. As $f'f = 2\ h$, and as there is a tensile stress of 80 in the diagonal $f'a'$ of the parallelogram $f'f\,a\,a'$, the compressive stress from f' to f and consequently from f' to the centre of the span is $80\,\dfrac{2\,h}{l} = 96$. In the same way the additional compressive stress from f to c is $40\,\dfrac{2\,h}{l} = 48$, and $48 + 96 = 144$ tons. Series HS.

Similarly for the *horizontal* tensile stress in the bottom table, as there is a compressive stress of 80 in the diagonal $f'a$ of the parallelogram $g\,f'a\,b$ the tensile stress from a to b, and consequently from a to c', is $80\,\dfrac{h}{l} = 48$; secondly, with the tensile stress of 80 in the diagonal $f'\,a'$ of the parallelogram $f'\,k'\,b\,a'$ there is a tensile stress from a' to c' of $80\,\dfrac{h}{l} = 48$; thirdly, with the compressive stress of 40 in the diagonal $f\,a'$ of the parallelogram $k'\,f\,a'\,b'$ there is a further tensile stress from a' to c' of 24, and $48 + 48 + 24 = 120$, the total stress between a' and e; lastly, the tensile stress of 40 in the diagonal $f\,e$ of the parallelogram $f\,k\,b'\,e$ will give a tensile stress of 24 from e to c' and $120 + 24 = 144$. Series HS′.

The stress therefore in the central bay of the bottom table is equal to that which has been found to exist in the two central bays of the top table.

Thus the horizontal stress at the centre of the span is less than 150 tons, or that which would be caused in a solid beam or in a plate girder under the same conditions of span, depth, and load.

This is owing to the fact that in the present example the diagonal and consequent horizontal action set up by vertical stress caused by the load does not commence at the centre of the span, but at the points e and e', each 3 feet distant from the centre.

---

122. A Rolling Load and a Single System Web. Referring to (**40, 109**) it will be seen that in the case of an evenly distributed load moving along a plate girder, when one-half of the span becomes covered by the load, the vertical stress at the centre of the span is one-fourth of the stress at each abutment due to the same load per unit of length when covering the whole span.

Now in the example Fig. 64, when the load covers half the span a to c' there will be as before a vertical stress from the load of 32 incident at e, and 32 also at a'. The proportionate amount of the stress at e passing in the direction e' to the further abutment will therefore be $\dfrac{32 \times 2}{5} = 12.8$, while that from a' passing in the same direction to the same abutment $= \dfrac{32 \times 1}{5} = 6.4$. Therefore the vertical stress at any point to the right of the vertical $k\,e$, and consequently that at the centre of the span, is $12.8 + 6.4 = 19.2$, being in this instance not quite one-fourth of that at each abutment when the whole span is covered by the load, because, as stated in the last Article, diagonal action in the web does not commence at the centre of the span.

In the following tabulated Series let A B, B C, C D, D E, and E F represent the five bays of the girder, and the dotted lines a, b, c, d, e the central verticals of each bay. Then when the load covers the span there are as has been seen 32 tons active at each vertical B, C, D, and E.

	A	a	B	b	C	c	D	d	E	e	F
VS $\{$			32 25.6 : 6.4		32 19.2 : 12.8		32 12.8 : 19.2		32 6.4 : 25.6		
	+	−	+	−	+	−	+	−	+	−	+	−
B′	32		(32 ......8)	8	8	8	8	8	8	8	
C′	24		24	24	(24 .....16)	16	16	16	16	16	
s	56		56	16	16	**24 24**	24	24	24	24	
D′	16		16	16	16	16	(16 ....24)	24	24	24	
s′	72		72	32	32	8	8	**48 48**	48	48	
E′	8		8	8	8	8	8	8	(8 .....32)	32	
S	80		80	40	40	0	0	40	40	**80 80**	

Assume the distributed rolling load to advance upon the girder from A towards F. Then when the advance end of the load is at *b* there will be a stress of 32 at B split (**30**) into two vertical stresses 25.6 and 6.4 as given Series VS. Now $25.6 \frac{l}{d} = 32$ is taken as a tensile stress by the diagonal tie-bar extending from B to *a* and as a compressive stress by the diagonal strut extending from *a* to A, while $6.4 \frac{l}{d} = 8$ is transmitted through a tie-bar and strut in each bay to F, as given Series B′, the sign + above the figures denoting compressive and the sign − tensile stress.

In the same way when the load has advanced to *c*, *d*, and *e* there will first be 32 incident at C, next 32 at D, and lastly 32 at E, the results due to each of these advances, when determined as before, being relatively tabulated in Series C′, D′, and E′.

The bracketed figures in each Series indicate the position of the fraction of the whole load to which that Series applies, and the figures in black type give the maximum stress in each web diagonal.

Series B', s, s', and S give the stress in each member of the web at each step in the advance of the load. These are found by addition if the terms are alike, and by subtracting the less amount from the greater if the terms are compressive and tensile, the result being a stress of the same kind as the greater amount or nil when the amounts are equal.

The last Series S is obviously that which would be given by the load when covering the whole span.

If the load were to advance upon the girder from right to left, the results given in Series B', C', D', and E', would simply be reversed from right to left. Thus it will be seen, Series s, that the two diagonals in the central bay C D should each be constructed so as to be capable of resisting either a thrust of 24 tons, or tensile stress of the same amount, whereas when the load extends over the whole span the stress in those members is nil. In the same way the strut and tie-bar in each of the bays B C and D E should be capable of resisting a stress of 48 tons, Series s', and not 40 tons only as given by Series S when the load covers the whole span.

Load Distributed.

123. Given a Double System Diagonal Framed Girder 30 feet long between the supports, 4 feet deep, and carrying at the bottom table a distributed load of 160 tons, or 5⅓ tons per foot run, Fig. 65. For the web, let the ratio of the lengths h, d, and l be 3, 4, and 5 (**115**).

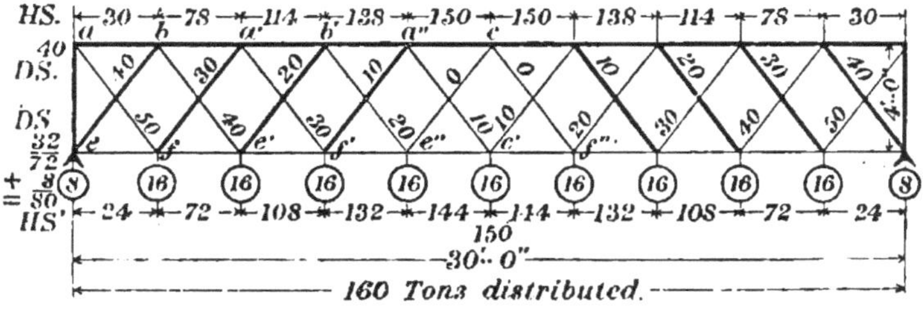

Fig. 65.

The length of the span is therefore divided into ten 3-feet bays as in example **117**, and the load being the same in both cases, a stress of $\dfrac{160}{10} = 16$ as indicated in the figure will become active at each angular junction of the web members with the bottom table, while the weight of one-half of the load carried by each outer bay will pass direct through the medium of the bottom table to the adjoining abutment.

As in example **119**, so also in this, the two systems of web members are quite distinct and separate. Reading from the centre of the span c' to the left support, the one system is $c'a''f'a'fa$, and the other $ce''b'e'be$.

To determine the stresses in the web members, each of these systems should therefore be separately treated. Let t be the total vertical stress at any given point as c' e'' f' in the line of the bottom table, and s the stress in a tie-bar. Then commencing with the first system at the centre c' of the span where $t = \frac{16}{2} = 8$, $s = \frac{tl}{d} = \frac{8 \times 5}{4} = 10$, or the tensile stress in the tie-bar $c'a''$, and consequently the thrust in the strut $a''f'$; this further causes a vertical stress of 8 at f', and therefore at that point $t = 8 + 16 = 24$, and $s = \frac{tl}{d} = \frac{24 \times 5}{4} = 30$, the tensile stress in the tie-bar $f'a'$, and consequently the thrust in the strut $a'f$. So also the tensile stress in the tie-bar af will be 50, and the vertical thrust in the end strut ae will be $50 \frac{d}{l} = 40$ tons. In the same way, commencing with the second system at e'', since the diagonals ce'' and cf'' are inactive, as has been shown (**121**), the stress of 16 at that point will cause a tensile stress of 20 in the tie-bar $b'e''$, and a thrust of 20 in the strut $b'e'$. The vertical stress at e' will thus be $16 + 16 = 32$, and the tensile stress in the tie-bar be' and the thrust in the strut be will be 40 tons.

It will be observed that the stress in a diagonal strut is not the same in amount as that in the tie-bar crossing it in any one of the ten bays.

The weight on the abutment brought by the strut be is therefore $40 \frac{d}{l} = 32$, to this add 40 caused by the tie-bar af through the vertical strut ae, and 8 coming direct from the bay $e\,f$ through the bottom table, then $32 + 40 + 8 = 80$, or half the total weight of the load.

The stress in each diagonal, as given, Series D S, D S, Fig. 65, having now been ascertained, the *horizontal* stress in each bay of the tables may, as in previous examples, be consecutively determined.

For the top table, commencing with the left-hand end bay, the horizontal stress in each bay will be as given in Series HS, Fig. 65.

$$\text{Thus} \quad . \quad . \quad . \quad . \quad . \quad 50\,\frac{h}{l} \;=\; 30 \qquad (1)$$

$$\left(40\,\frac{2\,h}{l} = 48\right) + \; 30 \;=\; 78 \qquad (2)$$

$$\left(30\,\frac{2\,h}{l} = 36\right) + \; 78 \;=\; 114 \qquad (3)$$

$$\left(20\,\frac{2\,h}{l} = 24\right) + 114 \;=\; 138 \qquad (4)$$

$$\left(10\,\frac{2\,h}{l} = 12\right) + 138 \;=\; 150 \qquad (5)$$

Equation (1) refers to the bay of the girder first given in which the diagonal $a\,f$ transfers a vertical stress · of $50\,\dfrac{d}{l} = 40$ to the end strut $a\,e$. Equation (2) extends over two bays defined by the diagonal $b\,e'$ of the parallelogram $b\,b'$, $e\,e'$, hence $2\,h$, and so on with the others following.

For the bottom table, commencing in the same way as before, the horizontal stress in each bay will be as given in Series HS′, Fig. 65.

$$\text{Thus} \quad . \quad . \quad . \quad . \quad . \quad 40\,\frac{h}{l} \;=\; 24 \qquad (1)$$

$$\left(50 + 30\,\frac{h}{l} = 48\right) + \; 24 \;=\; 72 \qquad (2)$$

$$\left(40 + 20\,\frac{h}{l} = 36\right) + \; 72 \;=\; 108 \qquad (3)$$

$$(30 \; + \; 10\,\frac{h}{l} = 24) \; + \; 108 \; = \; 132 \qquad (4)$$

$$(20 \; + \quad 0\,\frac{h}{l} = 12) \; + \; 132 \; = \; 144 \qquad (5)$$

$$. \; . \; . \; (10\,\frac{h}{l} = 6^*) \; + \; 144 \; = \; 150 \qquad (6)$$

Equation (1) clearly explains itself. The others when illustrated by equation (2) are formed in the following way.

There is a tensile stress of 50 meeting a thrust of 30 in the point *f*, and as *a f* and *a′ f* are diagonals of the two first bays or of two equal parallelograms these two factors may be added together, and thus combined in one operation. This method may also be applied to the top table, but the rationale of it would not be quite so apparent from the diagram at sight as that adopted.

Thus bay by bay, as with the diagonals of the web, so with the tables, the stresses are unequal.

124. A Rolling Load and a Double System Web.
The stress set up in each web member of the girder, Fig. 65, by the same load as given per length unit, when advancing on to, and upon the span, until the whole is covered, may be determined in the same way as in **122.**

In this example, however, two separate systems of web members necessitate the two following separate series of tabulated operations, in which let the line A F represent the length of the span divided into bays by vertical lines, that at (*c*) being the central vertical.

---

* This would be taken by a pin connecting the two tie-bars at *c′*, and not by the table itself, unless they had separate connections,

(1) *System terminating with Struts.*

	A	a	B	b	C	(c)	D	d	E	e	F

| VS. $\begin{cases} = \\ = \end{cases}$ | | | 16
12.8 : 3.2 | | 16
9.6 : 6.4 | | 16
6.4 : 9.6 | | 16
3.2 : 12.8 | | |

	+	−	+	−	+	−	+	−	+	−	+	−	+	−	+	−	+	−
B′	16		(16		4)	4		4	4		4	4			4	4		
C′	12		12	12		(12	8)	8	8		8	8			8	8		
s	$\overline{28}$		$\overline{28}$	8		8		$\overline{12}$	$\overline{12}$		$\overline{12}$	12			$\overline{12}$	$\overline{12}$		
D′	8		8	8		8 8			(8		12)	12			12	12		
s′	36		$\overline{36}$	$\overline{16}$		$\overline{16}$		4	4		$\overline{24}$	$\overline{24}$			$\overline{24}$	$\overline{24}$		
E′	4		4	4		4	4			4	4			(4	16)	16		
S	$\overline{40}$		$\overline{40}$	$\overline{20}$		$\overline{20}$	0		0		$\overline{20}$	20			$\overline{40}$	$\overline{40}$		

(2) *System terminating with Tie-bars.*

| VS. $\begin{cases} = \\ = \end{cases}$ | 16
14.4 : 1.6 | | 16
11.2 : 4.8 | | 16
8.0 : 8.0 | | 16
4.8 : 11.2 | | 16
1.6 : 14.4 |

	+	−	+	−	+	−	+	−	+	−	+	−	+	−	+	−	+	−
a′	(18		2)	2		2		2			2	2			2	2		2
b′	14	14		(14		6)	6		6	6		6	6		6	6		6
s′	$\overline{32}$	$\overline{12}$		$\overline{12}$		8	$\overline{8}$		$\overline{8}$	$\overline{8}$		$\overline{8}$	$\overline{8}$		$\overline{8}$	$\overline{8}$		$\overline{8}$
c′	10	10		10	10		(10		10)	18		10	10		10	10		10
s′	$\overline{42}$	$\overline{22}$		$\overline{22}$	$\overline{2}$		2		$\mathbf{18}$	$\mathbf{18}$		$\overline{18}$	$\overline{18}$		$\overline{18}$	$\overline{18}$		$\overline{18}$
d′	6	6		6	6		6	6		(6		14)	14		14			14
s″	$\overline{48}$	$\overline{28}$		$\overline{28}$	$\overline{8}$		$\overline{8}$		$\overline{12}$	12		$\mathbf{32}$	$\mathbf{32}$		$\overline{32}$			$\overline{32}$
e′	2	2		2	2		2	2			2	2		(2		18)		18)
S′	$\overline{50}$	$\overline{30}$		$\overline{30}$	$\overline{10}$		$\overline{10}$		10 10		$\overline{30}$	$\overline{30}$			$\overline{50}$			

Summaries of Results for Rolling Load.

(c)

(1) $\begin{cases} x \\ y \end{cases}$	$+16$	-16	-4	$+4$	-12	$+12$	-24	$+24$	-40	$+40$
x	$+16$	-16	-4	$+4$	-12	$+12$	-24	$+24$	-40	$+40$
y	$+40$	-40	$+24$	-24	$+12$	-12	$+4$	-4	-16	$+16$

(2) $\begin{cases} x \\ y \end{cases}$										
x	-18	-2	$+2$	-8	$+8$	-18	$+18$	-32	$+32$	-50
y	-50	$+32$	-32	$+18$	-18	$+8$	-8	$+2$	-2	-18

The relative sequence in the series of strains, Nos. (1) and (2) systems, assumes the advance of the load to be from left to right.

Thus as the load advances, a stress of 16 tons becomes incident in (1) system at each of the four verticals B, C, D, and E, and in (2) system at each of the five verticals a, b, (c), d, and e. Series VS give these increments of load stress, and the relative proportions into which each increment is split, at each point of incidence.

Each consecutive Series B′ C′, &c., is determined and relatively tabulated as stated (**122**).

The signs + and — relatively denote compressive and tensile stress, and the figures in black type denote that there is an increase of stress caused by the rolling load, as compared with that set up in the same members, by the same load per length unit, when covering the whole span.

In (1) and (2) summaries of results, Series x and y relatively give the stress in each web member during the advance of the load from left to right, and reversely from right to left.

The summary for (1) system shows that the rolling load subjects each of the two central web members to an alternate tensile and compressive stress of 12 tons, whereas when the load covers the whole span the stress in these members, is nil, Series S. Further, the stress in the two next members right and left remains the same in nature, but is raised from 20 to 24 tons.

The summary for (2) system shows that the two central web members are subjected to an alternate 8 tons compressive and 18 tons tensile stress, instead of 10 tons tensile stress when the load covers the span, as stated Series S′.

Note also the important increase of stress in the next

member right and left. This and other variations are seen in relative terms by comparing summary (2) with Series S' of (2) system.

Load Distributed.

125. To equalise the Stress in the like Parts of each Bay of a Double System Diagonal Framed Girder. Referring to Fig. 65, it will be seen that the stress in each bay of the top table is that in each collateral bay of the bottom table + 6, and that the stress in the tie-bar is that in the strut crossing the same bay + 10. The stress on the upper and lower connecting pins falling in the same vertical line will also in like manner vary. This is simply because, reading from the centre of the span, the load becomes active in one system, one bay in advance of the other.

To equate these systematic variations, let the span, depth, the arrangement of the web diagonals, and the load be the same as in the last example.

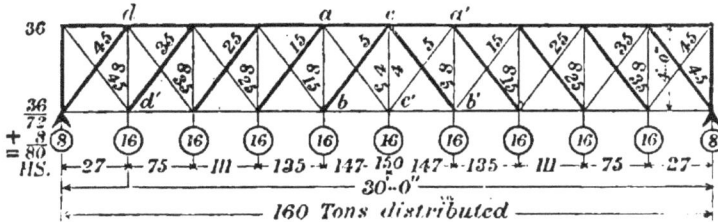

Fig. 66.

Then let a vertical tension bar* unite each upper angle of one system with each lower angle of the other in the same

* As in Sir John Hawkshaw's Charing Cross South Eastern Railway Bridge.

vertical line in which each fraction of the load, in this instance 16 tons, becomes active, Fig. 66.

Commencing at the foot c' of the central vertical, it will now be seen that of the 16 tons active at that point a stress of 8 will be taken by the two diagonal tie-bars c' a and c' a', while the remaining stress of 8 tons will pass through the tension bar c' c to the point c, there to be resisted by the two struts c b and c b'. Thus as the vertical stress taken by each of these four diagonals is $\frac{16}{4} = 4$, the stress in each will be $4 \frac{l}{d} = 5$.

Stress in the remaining diagonals may be determined as in previous examples, and as each additional increment of vertical stress is 16, so the additional stress in each successive pair of diagonals will be 10, for $\frac{16}{2} = 8$, and $8 \frac{l}{d} = 10$. Thus reading from the central vertical the series will now become 5, 15, 25, 35, 45 in either system as figured in the diagram.

By this arrangement, strain arising from any abnormal load-stress or shock occurring in any part of the span would be divided between the two systems by the introduction of the vertical tension bars, whereas without these the resistance to stress so caused might be principally thrown upon one system only. For, assuming that the tables have no vertical strength, should such a load-stress occur at any one of the points in which the web members are connected with either table, the stress caused by it would have to be resisted by one system only.

Stress in the two diagonals of each bay having been found to be equal, that in the collateral bays of each table will also be equal.

In determining the *horizontal* stress in the collateral bays of each table as given, Series HS, Fig. 65, commence as in the last article with the left-hand end bay.

Then . . . . $45\dfrac{h}{l} = 27$ (1)

$(80\,\dfrac{h}{l} = 48) + 27 = 75$ (2)

$(60\,\dfrac{h}{l} = 36) + 75 = 111$ (3)

$(40\,\dfrac{h}{l} = 24) + 111 = 135$ (4)

$(20\,\dfrac{h}{l} = 12) + 135 = 147$ (5)

$(\ 5\,\dfrac{h}{l} = 3*) + 147 = 150$ (6)

For Equation (2) the numeric factor is the stress in two diagonals acting at the same point d or d', or $45 + 35 = 80$, and so on with those which follow, excepting Equation (6) as one diagonal only is active at c or c' in the same direction.

It will be seen that diagonal and horizontal stress in each bay of the girder is now the mean of the results given in the last example.

Besides the additional efficiency already stated, there is yet another advantage attending this equation of stress in collateral members, for their constructive details would then precisely correspond each with each, and thereby constructive work would be simplified and facilitated.

But in girders of the type Fig. 65, this desirable arrange-ment of work could not be fairly carried out except in con-junction with an adequate provision for the series of major stresses in both systems.

---

* This stress is taken at the top table by the pin connecting the two struts, and at the bottom table by that connecting the two tie-bars, and not in either case by the table itself unless those web members should have separate connections.

Should the load be taken by the girder at the top instead of at the bottom table, the vertical members would then be struts, but without any alteration of the nature or amount of stress in any other member.

126. A Rolling Load and a Coupled Double System Web.

The stresses caused in the web members of the girder, Fig. 66, by the same load per length unit may be determined as stated (**122**) and (**124**), and are delineated in the following tabulated Series for the system terminating with struts.

Let the line A F of the Table represent the length of the span, divided into bays as in previous examples, (*c*) indicating the centre of the span, and let the load advance from left to right.

System terminating with Struts.

	A	a	B	b	C	(c)	D	d	E	e	F
vs. =		8	8	8	8	8	8	8	8	8	
vs. =		7.2:0.8	6.4:1.6	5.6:2.4	4.8:3.2	4:4	3.2:4.8	2.4:5.6	1.6:6.4	0.8:7.2	
	+ −	+ −	+ −	+ −	+ −	+ −	+ −	+ −	+ −	+ −	+ −
a'	(9	1)		1	1		1	1		1	1
B'	8	(8	2)	2	2	2	2	2	2	2	2
s₁	17	7		3	3		3	3		3	3
b'	7	7	(7	3)	3	3	3	3	3	3	3
s₂	24	14	4		**6**	6	6	6	6	6	6
C'	6	6	6		(6	4)	4	4	4	4	4
s₃	30	20	10		0	**10**	10	10	10	10	10
c'	5	5	5	5	(5	5)	5	5	5	5	5
s₄	35	25	15	5	5	**15**	15	15	15	15	15
D'	4	4	4	4	4		(4	6)	6	6	6
s₅	39	29	19	9	1	11	**21**	21	21	21	21
d'	3	3	3	3	3	3	(3	7)	7	7	7
s₆	42	32	22	12	2	8	18	**28**	28	28	28
E'	2	2	2	2	2	2	2		(2	8)	8
s₇	44	34	24	14	4	6	16	26	**36**	36	36
e'	1	1	1	1	1	1	1	1		(1	9)
S	45	35	25	15	5	5	15	25	35	**45**	

THE BEAM.

Summary of Results for Rolling Load.

(*c*)

x	$+ 9$	$+ 1$	$- 3$	$+ 6$	$- 10$	$+ 15$	$- 21$	$+ 28$	$- 36$	$+ 45$
y	$+ 45$	$- 36$	$- 28$	$- 21$	$+ 15$	$- 10$	$+ 6$	$- 3$	$+ 1$	$+ 9$

The signs $+$ and $-$ relatively denote compressive and tensile stress.

As the two systems of web members are now coupled, Series s_1, s_2, s_3, &c., relatively give the stress in each web member for each advance of the load bay by bay.

The figures in black type denote either increased stresses or a reverse in their nature or both when compared with those which obtain when the load covers the whole span.

In the summary of results, Series x and y relatively give the stress in each web member as the load advances from left to right and from right to left. Further, as stress in the two members, crossing each other in the same bay, has been equalised by the vertical tension-bars, so by reversing the signs $+$ and $-$ in this summary the two series x and y will give the stresses set up by the rolling load in the system terminating with tie-bars.

Comparing, therefore, as in the last example, Series x and y with Series S, the relative increase, and also the nature of the stress set up in each web member by the rolling load, is at once apparent.

127. Given a Quadruple System Lattice Girder, 30 feet long between the supports, 4 feet deep, and carrying at the bottom table a distributed load of 160 tons or 5⅓ tons per foot run. Figs. 67 and 68. For the web, let the ratio of the lengths *h, d,* and *l* be 3, 4, and 5 (**115**).

The length of the span is thus divided by the four systems of web members into twenty bays, and a vertical stress of $\frac{160}{20} = 8$ tons becomes active at each intermediate junction of the web members with the bottom table as indicated Fig. 68, while, as in the case of all framed and lattice girders, the weight of one-half of the distributed load carried by each outer bay, either at the top or the bottom table, in passing direct through the medium of the relative table to the adjoining abutment, will cause no strain in the web diagonals, or in the tables.

In this example each of the four systems of web members is theoretically quite distinct and separate.

Two of these four systems, Fig. 68, one carrying four and the other five of the nineteen equal fractions of the load start at two of the four extreme angles of the girder, and following the same lines as those in example Fig. 65 (**123**) terminate in like manner at the two extreme and opposite angles.

Thus the load at each point of incidence, being now half what it then was, the stresses in the members forming these two systems will be, as stated in the diagram, Fig. 68, half those given in that example.

The two remaining systems having peculiar charac-

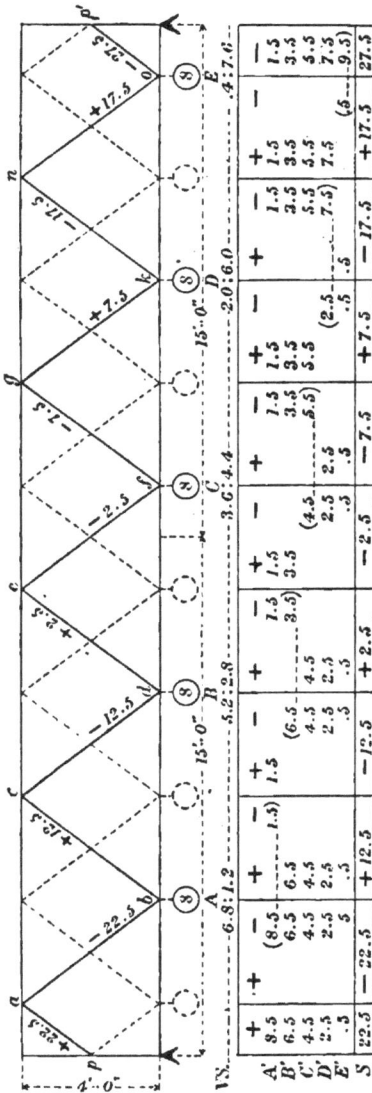

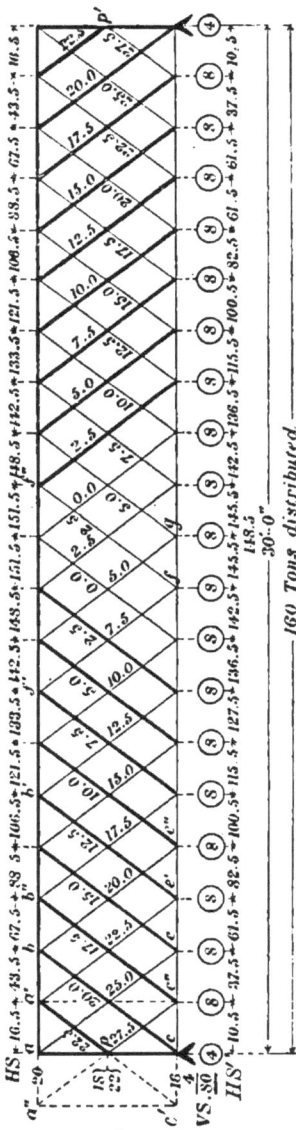

Fig. 67.

Fig. 68.

160 Tons distributed

teristics, are shown for convenience of treatment in the separate diagram, Fig. 67. These starting at a point, p, in one vertical end strut, midway in the depth of the girder, terminate at a similar point, p', in the other like and opposite strut, and being simply counterparts of each other *reversed* longitudinally, one of them for clearness of delineation is shown by dotted lines in the figure.

Each of these systems carries five of the nineteen fractions of 8 tons each, into which the active portion of the load is divided by the quadruple-system web. But as the points of incidence of these five fractions relative to the centre of the span and the points of support are not symmetrical, the determination of the stresses set up by them in the web diagonals of the system becomes a matter of special treatment.

First (**30**), find for one system the amount of vertical stress which travels from each of the five incidental fractions of 8 tons to each abutment through the medium of the web. These will be as given Fig. 67, Series VS. Thus commencing from the left; of the stress of 8 at A, 1.2 will travel to the right and 6.8 to the left support, and each such stress so obtained multiplied by $\dfrac{l}{d}$ will give that set up relatively right or left in the diagonals of the system by each incidental fraction of the load stress. For instance, from A to the right the stress in the diagonals will be $1.2\dfrac{l}{d} = 1.5$, while to the left it will be $6.8\,\dfrac{l}{d} = 8.5$, Series A', and so on for each incidental load stress at B, C, D, and E, the results will be as given in the corresponding Series B', C', D', and E'.

Throughout the Figure the sign + indicates compressive stress, and the sign -- tensile stress. When both tensile and compressive stress refer in the series to any one member, the amounts are so placed in two separate columns. Add up each of these columns, subtract the less aggregate sum from the greater, and the remainder will be the true stress either of compression or tension in the member, as given for each in the last Series S, and also relatively figured on the diagram.

For the counterpart system, taking each member in *reverse* longitudinal order as shown by dotted lines in the diagram and indicated by the letters A, B, C, D, E, above the tabulated Series (3) and (4), page 199 it will be at once evident that the stress in each relative diagonal of the two systems will be the same as that given for the first system in Series S, Fig. 67, and figured for both systems on the diagram Fig. 68.

The tabulated series of stresses, Fig. 67, gives intermediate results necessary for completing a determination of the maximum stresses to which the web members are subjected by a *rolling load* (**128**), but the following is a more simple method of determining the stresses in those members caused by an *evenly distributed load when covering the whole span.*

The five active portions of the load, Fig. 67, each = 8 being incident at distances from the right abutment of respectively 17, 13, 9, 5, and 1, twentieth parts of the span, the part of this load borne by the left abutment is $(17 + 13 + 9 + 5 + 1) \times \frac{8}{20} = 18$, thus causing an upward vertical reaction of 18 at p.

The effect of this will be a compressive stress in $p\,a$, amounting to $18 \times \frac{5}{4} = +22.5$, and a vertical stress acting upwards at a, equal to 18, causing a tensile stress in $a\,b$ of -22.5.

In the same manner, this occasions compressive stress of + 22.5 in $b\,c$, from which must be *deducted* the tensile stress due to the portion of the load which is active at b, or $8 \times \frac{5}{4} = 10$.

So that the resulting compressive stress ‾in $b\,c$ is $22.5 - 10 = + 12.5$.

In the same manner, it follows that the tensile stress in $c\,d$ is $- 12.5$.

$$
\begin{aligned}
\text{That in } d\,e \quad 12.5 - 10 &= +\ 2.5 \\
e\,f \quad . \quad . \quad . \quad . \ &= -\ 2.5 \\
f\,g \quad 2.5 - 10 &= -\ 7.5 \\
g\,k \quad . \quad . \quad . \quad . \ &= +\ 7.5 \\
k\,n \ -7.5 - 10 &= -\ 17.5 \\
n\,o \quad . \quad . \quad . \quad . \ &= +\ 17.5 \\
o\,p' -17.5 - 10 &= -\ 27.5
\end{aligned}
$$

As the stress in $e\,f$ is $- 2.5$, and that would cause a stress of $+ 2.5$ in $f\,g$, *add* $- 10$ and the result is $- 7.5$, and so on to $o\,p'$.

To verify the above, we may observe that the part of the load on this system carried by the right-hand abutment is $40 - 18 = 22$, and that $22 \times \frac{5}{4} = 27.5$ to which, being obviously a tensile stress owing to the direction of the member $o\,p'$, we give the negative sign and obtain $- 27.5$ in that member as before.

The stress in each diagonal member of the web having been ascertained, the stresses set up in each table may next be found (**123**), and then the results so obtained may be completed in the following way.

It will be observed that of the two unsymmetrical systems

of web members meeting in the points p, and p' the stress
at p is $+$ 22.5, and that at $p' -$ 27.5. Subtract independent
of signs the less from the greater, and the remainder will be 5.
Then $5\,\dfrac{h}{l} = 3$, a horizontal stress acting in the direction of
the centre of the span upon each end strut *midway* between the
two tables at the points p and p'. The effect of this stress (c)
upon each table will therefore be in this example $\dfrac{3}{2}$ or $1\frac{1}{2}$
tons. Therefore as the stress c acts in a reverse direction to the
tensile stresses in the bottom table as determined in accordance
with **123**, so $1\frac{1}{2}$ tons should be deducted from each of the
stresses thus given, and because it acts in the same direction
with the compressive stresses in the top table, so $1\frac{1}{2}$ tons should
be added to each of the stresses as given by each equation.

For the top table commencing with the left-hand end bay,
the horizontal stresses in the several bays will be as given in
Series HS, Fig. 68.

For the first bay, as $a\,b\,c\,e$ (Fig. 68), is a parallelogram, and
as the diagonal $a\,e$ equals 25 or the stress in that member, then
the resulting stress in $a\,b = 25\,\dfrac{h}{l} = 15$, and $15 + 1\frac{1}{2} = 16.5$.

For the second bay produce the diagonal $a'\,p$ to c', then as
$a'\,b'\,c'\,e'$ is a parallelogram of which the diagonal $a'\,e'$ equals
the side $a'\,c'$, and the stress in each of these members is
22.5; then $22.5\,\dfrac{2\,h}{l} = 27$, to this add 16.5 given by the
previous equation and the result will be 43.5, and so on for
each bay to the centre of the span. The value of the increment c having been added in the first, thus passes on in each
succeeding equation.

Thus . . . $(25\dfrac{h}{l} = 15) + c = 16.5$ (1)

$(22.5\dfrac{2\,h}{l} = 27) + 16.5 = 43.5$ (2)

$(20.0\dfrac{2\,h}{l} = 24) + 43.5 = 67.5$ (3)

$(17.5\dfrac{2\,h}{l} = 21) + 67.5 = 88.5$ (4)

$(15.0\dfrac{2\,h}{l} = 18) + 88.5 = 106.5$ (5)

$(12.5\dfrac{2\,h}{l} = 15) + 106.5 = 121.5$ (6)

$(10.0\dfrac{2\,h}{l} = 12) + 121.5 = 133.5$ (7)

$(7.5\dfrac{2\,h}{l} = 9) + 133.5 = 142.5$ (8)

$(5.0\dfrac{2\,h}{l} = 6) + 142.5 = 148.5$ (9)

$(2.5\dfrac{2\,h}{l} = 3) + 148.5 = 151.5$ (10)

For the bottom table, commencing in the same way as before, the horizontal stress in any bay will be as given in Series HS′, Fig. 68.

For the first bay it is evident that the thrust of 20 in the diagonal *b c* of the parallelogram *a b c e*, Fig. 68, will cause a tensile stress in the bottom table $= 20\dfrac{h}{l} = 12$, and this $- 1\frac{1}{2} = 10.5$.

For the second bay produce the line of the tie-bar *c″ p* to *a″*. Now there is a tensile stress of 27.5 acting in that bar in the direction of *a″ c″* upon the point *c″*, and there is also a compressive stress of 17.5 acting in the direction $b'' c''$ upon the same point, each of these tending to set up tensile stress in the bottom table from c'' to the centre of the span. Therefore as $a'' c''$ and $b'' c''$ are diagonals of two equal parallelograms, $a'' a' c' c''$ and $a' b'' c'' c'$, the two factors $27.5 + 17.5 = 45$ when thus combined cause a tensile stress in the bottom table from c'' to the centre of the span $= 45 \frac{h}{l} = 27$. To this add the stress given by Equation (1) and $27 + 10.5 = 37.5$, and so on with each succeeding bay to (9).

For the tenth bay, as the tensile stress $= 7.5$ in the diagonal $f f'$ would cause a tensile stress, and the tensile stress of 2.5 in the diagonal $f f''$ would cause a compressive stress in the bottom table from f to the centre of the span, and $7.5 - 2.5 = 5$, so $(5 \frac{h}{l} = 3) + 142.5 = 145.5$.

The stress of 3 tons given by Equation (11) will be taken by the table when each of the two tie-bars at g has a separate connection with it. But when there is but one connection, such as a bolt or pin, between the tie-bars themselves and the table, then the stress will be taken by the connection and not by the table.

$$\text{Thus . . } (20 \frac{h}{l} = 12) - c = 10.5 \tag{1}$$

$$(45 \frac{h}{l} = 27) + 10.5 = 37.5 \tag{2}$$

$$(40 \frac{h}{l} = 24) + 37.5 = 61.5 \tag{3}$$

$$(35 \frac{h}{l} = 21) + 61.5 = 82.5 \tag{4}$$

$$(30 \ \frac{h}{l} = 18) + 82.5 = 100.5 \tag{5}$$

$$(25 \ \frac{h}{l} = 15) + 100.5 = 115.5 \tag{6}$$

$$(20 \ \frac{h}{l} = 12) + 115.5 = 127.5 \tag{7}$$

$$(15 \ \frac{h}{l} = 9) + 127.5 = 136.5 \tag{8}$$

$$(10 \ \frac{h}{l} = 6) + 136.5 = 142.5 \tag{9}$$

$$7.5 - 2.5 = 5$$

$$\text{and } (5 \frac{h}{l} = 3) \quad + 142.5 = 145.5 \tag{10}$$

$$5 \frac{h}{l} = 3 * \ + 145.5 = 148.5 \tag{11}$$

The mean of the total central horizontal stresses in the top and bottom tables $= \dfrac{151.5 + 145.5 + 3\, *}{2} = 150$, or the stress caused in each table at the centre of the span with a continuous plate web under the same conditions of length, depth, and load.

The compressive stress in each vertical end strut is 20 from a to p and $20 + 18 + 22 = 60$ from p to c. For the stress at each abutment add 16 caused by the strut $b\, c$, and $\dfrac{8}{2} = 4$ caused by the fraction of load incident from c to c'', then the total stress is $60 + 16 + 4 = 80$, VS, Fig. 68.

128. A Rolling Load and a Quadruple System Lattice Girder, Figs. 67 and 68. Let the

---

* This increment of stress may or may not be taken by the bottom table **(117, 123)**.

line **A F** and each like horizontal line in the following tabulated Series represent the length of the span divided into bays by vertical lines, that at (*c*) being the central vertical, and let the same load as before advance from either end upon the girder until the whole is covered.

(1) *System terminating with full-length Struts.*

(1)		A	a	B	b	C	(c)	D	d	E	e	F
	VS			8 / 6.4 : 1.6		8 / 4.8 : 3.2		8 / 3.2 : 4.8		8 / 1.6 : 6.4		
	S	+20	−20	+10	−10	0	0	−10	+10	−20	+20	

Summary of Results for Rolling Load.

		A	a	B	b	C	(c)	D	d	E	e	F
	x	+8	−8	−2	+2	−6	+6	−12	+12	−20	+20	
	y	+20	−20	+12	−12	+6	−6	+2	−2	−8	+8	

(2) *System terminating with full-length Tie-bars.*

(2)		A	a	B	b	C	(c)	D	d	E	e	F
	VS		8 / 7.2 : 0.8		8 / 5.6 : 2.4			8 / 4.0 : 4.0		8 / 2.4 : 5.6		8 / 0.8 : 7.2	
	S	−25	+15	−15	+5	−5	−5	+5	−15	+15	−25	

Summary of Results for Rolling Load.

		A	a	B	b	C	(c)	D	d	E	e	F
	x	−9	−1	+1	−4	+4	−9	+9	−16	+16	−25	
	y	−25	+16	−16	+9	−9	+4	−4	+1	−1	−9	

(3) *System commencing with a half-length Strut at p, Fig.* 67.

	A	B	C	D	E
VS	$\dfrac{8}{6.8:1.2}$	$\dfrac{8}{5.2:2.8}$	$\dfrac{8}{3.6:4.4}$	$\dfrac{8}{2.0:6.0}$	$\dfrac{8}{0.4:7.6}$
S	+22.5 −22.5	+12.5 −12.5	+ 2.5 − 2.5	− 7.5 + 7.5	−17.5 +17.5 −27.5

Summary of Results for Rolling Load.

	A	B	C	D	E
x	+ 8.5 − 8.5	− 1.5 + 1.5	− 5.0 + 5.0	−10.5 +10.5	− 18.0 +18.0 − 27.5
y	+22.5 −22.5	+14.0 −14.0	+ 7.5 − 7.5	+ 3.0 − 3.0	− 7.0 − 0.5 − 9.5

(4) *System commencing with a half-length Tie-bar at p, Fig.* 67.

	E	D	C	B	A
VS	$\dfrac{8}{7.6:0.4}$	$\dfrac{8}{6.0:2.0}$	$\dfrac{8}{4.4:3.6}$	$\dfrac{8}{2.8:5.2}$	$\dfrac{8}{1.2:6.8}$
S	−27.5 +17.5	−17.5 + 7.5	− 7.5 − 2.5	+ 2.5 −12.5	+12.5 − 22.5 +22.5

Summary of Results for Rolling Load.

	E	D	C	B	A
x	− 9.5 − 0.5	− 7.0 − 3.0	+ 3.0 − 7.5	+ 7.5 −14.0	+14.0 −22.5 +22.5
y	−27.5 +18.0	−18.0 +10.5	−10.5 + 5.0	− 5.0 + 1.5	− 1.5 − 8.5 + 8.5

It has been shown (**127**) that the vertical stress caused by the load at each point of incidence is 8 tons. Therefore the stresses given in the Series for (1) and (2) systems of web members are one-half of those determined (**123, 124**) for the like systems and tabulated, page 182. For as there are now twice the number of web systems, so there is half the vertical stress at each point of incidence caused by the same load per foot run, and consequently half the stress in each web member. To facilitate reference the verticals of these two systems are lettered relatively with those (**124**).

The stresses for (3) system of web members are deductions from the Series A′, B′, C′, D′, and E′, Fig. 67 (**127**).

The stresses for (4) system are those determined for (3) system reversed end for end of the girder.

Series S give for each of the four systems respectively the stress in tons in each web member when the whole span is covered by the load.

Series x and y give for each system the stress in tons in each web member during the advance of the load from left to right and from right to left.

The signs + and − respectively denote compressive and tensile stress.

Upon examination of the whole of the series it will be seen that the greatest tensile stress in a central member is 9 tons and in an end member $27\frac{1}{2}$ tons, also that the greatest compressive stress in a central member is 6 tons, and $22\frac{1}{2}$ tons in an end member.

The stresses set up by a rolling load in each of the web members of the lattice girder, Fig. 68, have thus been theoretically determined. The spans of such girders, and the distributed loads they are intended to carry, are, however, so various that the relative sections of the web members must in some measure be determined in accordance with practical requirements. See (**106**) and (**109**) on plate webs.

For instance, in the treatment of small girders intended to carry light loads, the section of a member or the part of a member which would have to resist the greatest stress having been determined, that section might, in order to simplify work, be judiciously adopted throughout the like members.

Thus in ordinary practice, the final relative proportions of the members of a structure are frequently best adjusted in accordance with the results of practical knowledge and experience.

129. The depths suitable for framed and lattice girders as arranged in the following Table were computed, like those for plate girders Table 8 (**103**), from the diagram given by Sir Benjamin Baker.* In the same way as with the former table it extends from girders of 20 to 200 feet span, and from a load of 10 to 100 cwt. per foot run, thus giving for reference data covering a wide range of conditions which may have to be met in ordinary girder work.

TABLE 9.

DEPTHS FOR LATTICE GIRDERS.

Feet span.	20		40		60		80		100		120		140		160		180		200	
Per ft. run. cwt.	Ft.	ins.	Ft.	ins.	Ft.	ins.	Ft.	ins.	Ft.	ins.	Ft.	ins.	Ft.	ins.	Ft.	ins.	Ft.	ins.	Ft.	ins.
10	2	5	4	9	7	0	9	1	11	2	13	0	14	11	16	8	18	4	20	0
20	2	7	5	1	7	5	9	8	11	11	14	0	16	0	18	0	19	11	21	9
30	2	9	5	4	7	9	10	2	12	6	14	10	16	10	19	0	21	1	23	0
40	2	10	5	7	8	1	10	7	13	1	15	6	17	8	20	1	22	1	24	2
50	2	11	5	9	8	5	11	0	13	7	16	1	18	6	20	11	23	1	25	3
60	3	0	5	11	8	8	11	4	14	1	16	8	19	2	21	8	24	0	26	4
70	3	1	6	1	8	11	11	8	14	6	17	2	19	10	22	4	24	11	27	4
80	3	1	6	2	9	1	12	0	14	11	17	8	20	4	23	0	25	8	28	2
90	3	2	6	3	9	3	12	3	15	3	18	0	20	10	23	7	26	4	29	0
100	3	2	6	4	9	4	12	5	15	6	18	4	21	2	24	0	26	10	29	6

* "The Strength of Beams, Columns, and Arches." B. Baker, 1870.

The extreme variations in the ratio of depth to span for plate and lattice girders as given in Tables 8 and 9 are with s as the span and d the depth.

Plate
$$\begin{cases} 200 \text{ feet span with load } 10 \text{ cwt. per foot } \quad d = \dfrac{s}{14} \\[2ex] 20 \quad ,, \qquad ,, \quad 100 \quad ,, \qquad ,, \qquad d = \dfrac{s}{5.6} \end{cases}$$

Lattice
$$\begin{cases} 200 \quad ,, \qquad ,, \quad 10 \quad ,, \qquad ,, \qquad d = \dfrac{s}{10} \\[2ex] 20 \quad ,, \qquad ,, \quad 100 \quad ,, \qquad ,, \qquad d = \dfrac{s}{6.3} \end{cases}$$

CHAPTER XI.

130. The Modulus of Elasticity is the measure within the limits of perfect elasticity of the force required to produce a certain amount of extension or compression in a body in any one direction, as for instance in the length of an iron bar when subjected longitudinally to tensile or compressive stress.

Now experiments and tests have shown that within such limits each additional unit of stress will cause an additional equal unit of elongation or compression. Thus, it is found that a tensile stress of one pound applied longitudinally will extend a cast-iron bar one square inch in section $\frac{1}{17,000,000}$ part of its length, therefore 17,000,000 pounds would be required to extend the same bar to twice its original length, if that were possible within the limits of perfect elasticity, and E, the symbol of the modulus of elasticity = 17,000,000, while for wrought-iron it is 24,000,000.*

Tables of the modulus of elasticity for materials may be found in treatises which embrace the theory of deflection.

The deflection of a beam when transversely strained within the limits of perfect elasticity is (1) directly as the load or stress; (2) as the cube of the span, because the horizontal

* Hodgkinson. For different samples, however, the value of E will vary considerably.

strain with a given load varies directly as the length, while the versed sine of a comparatively short segment of an arc of large radius varies as the square of the length; (3) and for the same reason inversely as the cube of the depth; (4) also inversely as the breadth and as the modulus of elasticity.

The deflection of a solid rectangular beam carrying a central load is represented by the following equation :

Let l = the length . . . . in inches.
  b breadth . . . .   ,,
  d depth . .   ,,
  W   central load . . .   pounds.
  E   modulus of elasticity
  δ deflection . . .   inches.

$$\text{Then } \delta = \frac{W\, l^3}{4\, E\, b\, d^3} \qquad (27.)$$

This equation will apply only so long as the limits of elasticity are not exceeded.

131. The Deflection of *solid beams* and *flanged girders* may under the following conditions of load or stress be readily found by equation (28), in which c is a *coefficient* determined by actual experimental tests,* l the length in feet, d the depth or diameter, and δ the deflection in inches.

$$\delta = \frac{c\, l^2}{d} \qquad (28.)$$

The deflection caused by a *distributed load* will be $\frac{5}{8}$ths of that due to an equal *central load*, and the deflection caused by a *distributed load* of any fraction of the breaking weight will

---

* " Beams, Columns, and Arches." Sir Benjamin Baker.

be one fourth more than that due to the same fraction of the *central* breaking load.

Coefficients for *solid beams* under a *central load* of one-fourth the breaking weight.

Section of Beam.	Cast Iron.*	Wrought Iron.	Steel.
Rectangular	.017	.019	.026
Round	.019	.021	.029

Coefficients for *uniformly loaded girders* under tensional stresses in the bottom table of 1½ tons, 4½ tons, and 6½ tons for cast iron, wrought iron, and steel respectively.

Form of Girder.	Cast Iron.*	Wrought Iron.	Steel.
Uniform depth and section	.0192	.0135	.0156
Depth and section reduced from the centre of span to each end }	.0228	.0162	.0185

Coefficients for girders intended to carry a *distributed load* under tensile *proof stresses* in the bottom table of 2½ tons, 5 tons, and 8 tons for cast iron, wrought iron, and steel respectively.

Form of Girder.	Cast Iron.*	Wrought Iron.	Steel.
Uniform depth and section	.0264	.012	.01596
Depth and section reduced from the centre of span to each end }	.0336	.015	.01992

* The area of the bottom table being four times that of the top table.

Should the deflection of a girder on being tested advance more rapidly than expected, and to a greater extent than that given by careful calculation, the efficiency of the material or the soundness of the structure would be doubtful.

INDEX.

PRINTED BY NICHOLS & SONS, 25, PARLIAMENT STREET, S.W.